Springer Series on Polymer and Composite Materials

Series Editor

Susheel Kalia, Army Cadet College Wing, Indian Military Academy, Dehradun, India

Editorial Board

Kazutoshi Haraguchi, Department of Applied Molecular Chemistry, Nihon University, Narashino, Japan

Annamaria Celli, Department of Civil, Chemical, Environmental, and Materials Engineering, University of Bologna, Bologna, Italy

Eduardo Ruiz-Hitzky, Instituto de Ciencia de Materiales de Madrid, Spanish National Research Council, Madrid, Spain

Alexander Bismarck, Department of Chemical Engineering, Imperial College London, London, UK

Sabu Thomas, School of Chemical Sciences, Mahatma Gandhi University, Kottayam, Kerala, India

Michael R. Kessler, College of Engineering, North Dakota State University, Fargo, USA

Balbir Singh Kaith, Department of Chemistry, Dr B.R. Ambedkar National Institute of Technology, Jalandhar, India

Luc Averous, University of Strasbourg, Strasbourg, France

Bhuvanesh Gupta, Department of Textile Technology, Indian Institute of Technology Delhi, New Delhi, Delhi, India

James Njuguna, School of Engineering, Robert Gordon University, Aberdeen, UK

Sami Boufi, University of Sfax, Sfax, Tunisia

Magdy W. Sabaa, Faculty of Science, Cairo University, Cairo, Egypt

Ajay Kumar Mishra, College of Science, University of South Africa, Johannesburg, South Africa

Krzysztof Pielichowski, Department of Chemistry and Technology of Polymers, Cracow University of Technology, Krakow, Poland

Youssef Habibi, Materials Research and Technology Department, Luxembourg Institute of Science and Technology, Esch-sur-Alzette, Luxembourg

Maria Letizia Focarete, Department of Chemistry G. Ciamician, University of Bologna, Bologna, Italy

Mohammad Jawaid, Biocomposite Technology Laboratory, Universiti Putra Malaysia, Serdang, Malaysia

The "Springer Series on Polymer and Composite Materials" publishes monographs and edited works in the areas of Polymer Science and Composite Materials. These compound classes form the basis for the development of many new materials for various applications. The series covers biomaterials, nanomaterials, polymeric nanofibers, and electrospun materials, polymer hybrids, composite materials from macro- to nano-scale, and many more; from fundamentals, over the synthesis and development of the new materials, to their applications. The authored or edited books in this series address researchers and professionals, academic and industrial chemists involved in the areas of Polymer Science and the development of new Materials. They cover aspects such as the chemistry, physics, characterization, and material science of Polymers, and Polymer and Composite Materials. The books in this series can serve a growing demand for concise and comprehensive treatments of specific topics in this rapidly growing field. The series will be interesting for researchers working in this field and cover the latest advances in polymers and composite materials. Potential topics include, but are not limited to:

Fibers and Polymers:

- Lignocellulosic biomass and natural fibers
- Polymer nanofibers
- Polysaccharides and their derivatives
- Conducting polymers
- Surface functionalization of polymers
- Bio-inspired and stimuli-responsive polymers
- Shape-memory and self-healing polymers
- Hydrogels
- Rubber
- Polymeric foams
- Biodegradation and recycling of polymers

Bio- and Nano- Composites:-

- Fiber-reinforced composites including both long and short fibers
- Wood-based composites
- Polymer blends
- Hybrid materials (organic-inorganic)
- Nanocomposite hydrogels
- Mechanical behavior of composites
- The Interface and Interphase in polymer composites
- Biodegradation and recycling of polymer composites
- Applications of composite materials

Peter Rantuch

Ignition of Polymers

Springer

Peter Rantuch
Faculty of Materials Science
and Technology in Trnava
Slovak University of Technology
in Bratislava
Trnava, Slovakia

ISSN 2364-1878　　　　　　ISSN 2364-1886　(electronic)
Springer Series on Polymer and Composite Materials
ISBN 978-3-031-13081-6　　　ISBN 978-3-031-13082-3　(eBook)
https://doi.org/10.1007/978-3-031-13082-3

© The Editor(s) (if applicable) and The Author(s), under exclusive license to Springer Nature Switzerland AG 2022

This work is subject to copyright. All rights are solely and exclusively licensed by the Publisher, whether the whole or part of the material is concerned, specifically the rights of translation, reprinting, reuse of illustrations, recitation, broadcasting, reproduction on microfilms or in any other physical way, and transmission or information storage and retrieval, electronic adaptation, computer software, or by similar or dissimilar methodology now known or hereafter developed.

The use of general descriptive names, registered names, trademarks, service marks, etc. in this publication does not imply, even in the absence of a specific statement, that such names are exempt from the relevant protective laws and regulations and therefore free for general use.

The publisher, the authors, and the editors are safe to assume that the advice and information in this book are believed to be true and accurate at the date of publication. Neither the publisher nor the authors or the editors give a warranty, expressed or implied, with respect to the material contained herein or for any errors or omissions that may have been made. The publisher remains neutral with regard to jurisdictional claims in published maps and institutional affiliations.

This Springer imprint is published by the registered company Springer Nature Switzerland AG
The registered company address is: Gewerbestrasse 11, 6330 Cham, Switzerland

Preface

In general, the ignition of a material may be characterised as the initial point between the state when it is not burning and the state when it is burning. Combustion can be either homogeneous or heterogeneous. While homogeneous combustion occurs in gaseous phases (and is characterised by the occurrence of a flame), heterogeneous combustion occurs on the borderline of two phases and takes the form of smoke accompanied by luminescence of the solid surface (smouldering, glowing). Flame combustion tends to start suddenly, allowing us to quite accurately identify the moment of initiation. It is characterised by the initiation characteristics, through which we can not only assume the time to ignition but also predict its value if the external conditions change. At the same time, they can be used to describe different materials and compare them in terms of their flammability. Although these characteristics are applied most often in the area of fire protection, they may also be used within energetics or industrial processing.

Polymer materials are widely used in practice. This is partially due to the fact that this group includes a substantial range of material. But what is more, many of them provide excellent possibilities for the production of various macro-, micro-, or even nano-composites. Most polymers are flammable. This is why we need to understand their behaviour when they are exposed to an external heat flux. The following pages will therefore describe the basic information related to thermal decomposition, the effect of an incident heat flux on initiation time, methods for the calculation of initiation parameters, and their values for selected polymers. I believe that this publication will not only be helpful to professionals dealing with this topic, but also to beginners who wish to understand the process behind the initiation of burning and its characterisation.

Trnava, Slovakia   Peter Rantuch

Acknowledgements This work was supported by the Slovak Research and Development Agency under the contract No. APVV-16-0223. This work was also supported by the KEGA agency under the contracts No. KEGA 016STU-4/2021 and KEGA 001TU Z-4/2020. This work was also supported by the VEGA agency under the contracts No. VEGA 1/0678/22.

Special thanks go to Plasty Mladeč, which provided PM filaments for the research part of this publication.

Contents

1	**The Thermal Degradation of Polymer Materials**		1
	1.1 Polymers and Their Composition		1
	1.2 Additives Added to Polymers		2
		1.2.1 Lubricants	4
		1.2.2 Plasticisers	5
		1.2.3 Colourants	5
		1.2.4 Fillers	5
		1.2.5 Stabilisers	6
		1.2.6 Blowing Agents	7
		1.2.7 Fire Retardants	8
	1.3 The Impact of Increased Temperature on Polymers		10
	1.4 The Flammability of Polymers		12
	1.5 The Thermal Degradation of Selected Polymers		16
		1.5.1 Polypropylene (PP)	16
		1.5.2 Polyethylene (PE)	17
		1.5.3 Polyvinylchloride (PVC)	19
		1.5.4 Polyurethanes (PUR)	21
		1.5.5 Polystyrene (PS)	22
		1.5.6 Polylactic Acid (PLA)	23
		1.5.7 Acrylonitrile Butadiene Styrene Copolymer (ABS)	25
		1.5.8 Polyethylene Terephthalate (PET) and Polyethylene Terephthalate Glycol (PETG)	27
		1.5.9 Natural Polymers	29
	References		38
2	**The Correlation Between External Heat Flux and Time to Ignition**		45
	2.1 General Assumptions		45
	2.2 Suggested Correlations		52
		2.2.1 Determining Correlation According to Lawson and Simms [8]	52

	2.2.2	Determining Correlation According to Koohyar [9], Hallman [1] and Wesson et al. [2]	53
	2.2.3	Determining Correlation According to Smith and Satija [12]	55
	2.2.4	Determining Correlation According to Quintiere and Harkleroad [14]	56
	2.2.5	Determining Correlation According to Bluhme [16]	56
	2.2.6	Determining Correlation According to Mikkola and Wichman [18]	57
	2.2.7	Determining Correlation According to Delichatsios et al. [20]	60
	2.2.8	Determining Correlation According to Janssens [7]	61
	2.2.9	Determining Correlation According to Spearpoint and Quintiere [21]	62
	2.2.10	Determining Correlation According to Harada [22]	63
	2.2.11	Determining Correlation According to Shi and Chew [23]	65
	2.2.12	Determining Correlation According to Babrauskas [24]	65
	2.2.13	Determining Correlation According to An et al. [25]	66
	References		67
3	**Methods of Calculation of Ignition Parameters**		**69**
	3.1	The Most Frequently Used Apparatus for the Measurement of Time to Ignition	69
		3.1.1 Ignition Cabinet	69
		3.1.2 Lateral Flame Spread Apparatus	70
		3.1.3 Fire Propagation Apparatus	71
		3.1.4 Ignitability Test Apparatus	73
		3.1.5 Cone Calorimeter	73
	3.2	The Calculation of Ignition Parameters	75
		3.2.1 Critical Heat Flux	75
		3.2.2 Flux–Time Product	78
		3.2.3 The Thermal Response Parameter	78
		3.2.4 The Transfer Convective Coefficient	79
		3.2.5 The Apparent Thermal Inertia and Thermal Diffusivity	79
		3.2.6 The Ignition Temperature	82
	References		84
4	**Comparing the Ignition Parameters of Various Polymers**		**87**
	4.1	Materials	87
		4.1.1 Plastics	87
		4.1.2 Wood-Based Materials	88
	4.2	Methods Used to Compare the Ignition Parameters of Different Polymers	90

4.3	Results		91
	4.3.1	Plastics	91
	4.3.2	Wood-Based Materials	98
4.4	Discussion		102
	4.4.1	Synthetic Polymers	104
	4.4.2	Natural Polymer Materials	110
References			111

Conclusion 115

Symbols

α_{abs}	Absorptivity
$\bar{\alpha}$	Average absorptivity
β_{ig}	Ratio of convective gain and radiative loss with incident heat flux at ignition
ΔT	The change in temperature
δ	Thermal penetration depth
ε	Emissivity
κ	Thermal diffusivity
κ_T	Thermal diffusivity at temperature T
π	Ludolf's number
ρ	Density
ρ_d	Volume weight in a dry state
ρ_R	Reflectivity
ρ_T	Density at temperature T
σ	Stefan–Boltzmann constant
τ	Non-dimensional time
Φ	Constant
Ψ	Heat loss rate per unit area for one degree rise in temperature
A	Constant for pilot ignition
a	Constant
a_1	Constant
a_2	Constant
a_3	Constant
a_4	Constant
a_5	Constant
a_c	Empirical constant
A_δ	Coefficient for thermal penetration depth calculation
B	Constant for autoignition
b_4	Constant
b_c	Empirical constant
B_i	Biot number

c_c	Empirical constant
c_p	Thermal capacity
c_T	Thermal capacity at temperature T
erf	Error function
FTP	Flux–time product
H_a	Air humidity
h	Heat loss coefficient
h_c	Convective heat transfer coefficient
h_{ig}	Convection coefficient at ignition
I	Thermal inertia
$ierfc$	First integral of the complementary error function
b	Constant
K	Coefficient of thermal conductivity
$K_\perp$	Thermal conductivity in the perpendicular direction
$K_\parallel$	Thermal conductivity in the parallel direction
K_T	Thermal conductivity at temperature T
K_{T_g}	Thermal conductivity at glass transition temperature
L	Thickness
L_0	Sample thickness
L_s	Thickness of the surface layer
L_v	Heat of gasification
M	Moisture
m_{cr}	Mass flux of fuel vapours
n_{por}	Porosity
p_{atm}	Atmospheric pressure
q	Heat flux
q_a	Reflected external heat flux
q_b	External heat flux heating surface layer
q_c	External heat flux reaching the deeper layers
q_{cr}	Critical heat flux
$q_{cr(H)}$	Critical heat flux in the horizontal position
$q_{cr(V)}$	Critical heat flux in the vertical position
q_e	External heat flux reaching
q_i	Incident heat flux
q_{in}	Sum of total entering heat fluxes
$q_{intercept}$	Intercept heat flux
q_o	Overall heat flux
q_n	Net heat flux
q_{out}	Sum of total leaving heat fluxes
q_{loss}	Surface heat loss
q_{loss_r}	Heat loss through re-radiation
q_{loss_c}	Heat loss by convection
SPF	Self-propagation flux
T	Thermodynamic temperature
t	Time

Symbols

T_0	Ambient temperature
T_{bb}	Temperature corresponding to a black body
T_g	Glass transition temperature
T_{grad}	Temperature gradient
T_{ig}	Ignition temperature
t_{ig}	Time to ignition
T_{max}	Maximum surface temperature
T_p	Surface temperature at ignition
T_r	Temperature of the rear face
T_s	Surface temperature
TRP	Thermal response parameter
y	Spatial coordinate
z	Constant

Chapter 1
The Thermal Degradation of Polymer Materials

1.1 Polymers and Their Composition

Polymers are macromolecular substances with molecules composed of monomer units. These are interconnected by chemical bonds that create chain or cross-linked structures. If a polymer is composed of units of a single monomer, it is called a homopolymer. If its structure is composed of two or more types of monomers, it is called a copolymer. For instance, a monomer unit of styrene may create a homopolymer polystyrene but also a number of copolymers, such as acrylonitrili butadiene styrene (ABS). The process of the formation of a polymer macromolecule from monomer units is called polymerisation.

Polymers are divided into synthetic and natural polymers based on their origin. Typical examples of synthetic polymers include polyethylene, polyvinylchloride, and others. They are often described using abbreviations (Table 1.1). Natural polymers include proteins or cellulose. Ducháček [1] divides polymers into elastomers and plastics, and in the case of the latter, he further distinguishes between thermoplastics and reactoplastics. He characterises elastomers as highly elastic polymers that, under normal circumstances, may be significantly deformed using a mild degree of force without causing any damage and any deformation tends to be reversible. He continues by describing plastics as polymers that tend to be hard and often fragile under normal conditions. When exposed to higher temperatures, they become plastic and formable. If this change of state is repeatable, they are classed as thermoplastics, if it is irreversible, they are classed as reactoplastics. Although this classification is quite common in literature, we must bear in mind that this classification is related to the processing synthetic polymers. Hence, it does not apply to natural polymers.

Although there are also several inorganic materials that could be classified as polymers, they are mostly organic materials, which means that their structure is composed of atoms of carbon. Furthermore, they often include elements such as hydrogen, oxygen, nitrogen, and others.

Table 1.1 Abbreviations of selected synthetic polymers

Abbreviation	Polymer	Abbreviation	Polymer
PVC	Polyvinyl chloride	PES	Polyethersulphone
PP	Polypropylene	PMMA	Polymethyl methacrylate
PE	Polyethylene	ABS	Acrylonitrile butadiene styrene copolymer
PET	Polyethylene terephthalate	PMO	Polyoxymethylene
PAN	Polyacrylonitrile	PUR	Polyurethane
PS	Polystyrene	SAN	Styrene acrylonitrile copolymer
PTFE	Polytetrafluoroethylene	EVA	Ethylene–vinyl acetate copolymer
PA	Polyamide	SI	Silicone
PB	Poly-1-butene	PBT	Polybutylene terephthalate
PC	Polycarbonate	PLA	Polylactic acid
PUR	Polyurethane	PA	Polyamide
PAI	Polyamidoamine	PB	Polybutene
PBD	1,2-polybutadiene	PC	Polycarbonate
PETG	Polyethylene terephthalate glycol	PBI	Polybenzimidazole

The properties of polymer materials often depend on several factors that, in addition to their chemical composition, include the conditions and methods of polymerisation, molecular weight, and the shape and spatial organisation of the macromolecules [2].

Based on the demand for plastics within Europe (Fig. 1.1), it is obvious that the plastics that have long been in the greatest demand are polypropylene, polyethylene, and polyvinylchloride [3–12]. To maintain the objective nature of the data illustrated in the chart in Fig. 1.1, we should clarify that until 2013, it included data for the EU member states, plus Norway and Switzerland. Since 2014, it has also included data from Croatia, which, however, does not play a major role with regard to demand for plastics in Europe.

1.2 Additives Added to Polymers

The properties of polymer materials tend to be modified by the addition of various additives to meet practical requirements. These are not typically chemically bound with the macromolecule, but are attached to it through physical forces. This process creates composite materials, and the matrix of which is a polymer. In order to achieve the maximum degree of homogeneity of the resulting mixture, it is necessary for the additive to have the smallest possible particle size; hence, they are often used in a

1.2 Additives Added to Polymers

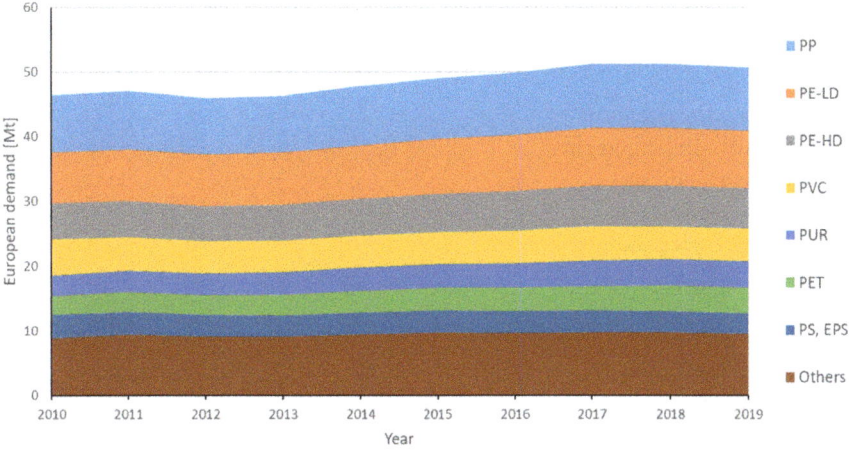

Fig. 1.1 European plastic demand (according to Plastics Europe [3–12])

powder form or as small fibres. If these additives are added to a melted polymer, their mutual miscibility is highly important. If the additive does not mix well with the polymer, its particles are modified on the surface. In this way, we are able to significantly expand the use of possible additives for a wide range of polymers.

However, apart from the required changes, additives also bring the risk that they might affect other properties of the resultant material. These changes may be considered negative for a given application or even make it impossible to use the polymer composite. For this reason, it is crucial to use the correct mixing ratio for the additives.

Weight/weight percentage (w/w) is used to specify the amount of additive in a polymer material, but in certain cases it is specified using phr (parts per hundred resin) that is the required weight of the additive that should be added to 100 weight units of polymer (phr).

Considering the need to modify plastics for a large number of applications, there is a wide range of additives. Based on the properties that they modify, we may classify them as follows [13]:

1. Surface Property Modifiers
   (a) Antiblocking agents
   (b) Antifogging agents
   (c) Antistatic agents
   (d) Coupling agents
   (e) Release agents.

2. Chemical Property Modifiers
   (a) Antioxidants
   (b) Biocides
   (c) Flame retardants

(d) Ultraviolet stabilisers.

3. Processing Modifiers

   (a) Blowing agents
   (b) Cross-linking agents
   (c) Heat stabilisers
   (d) Plasticisers
   (e) Processing aids
   (f) Lubricants.

4. Mechanical Property Modifiers

   (a) Fillers
   (b) Impact modifiers
   (c) Nucleating agents
   (d) Reinforcing fibres.

5. Aesthetic Property Modifiers

   (a) Colouring agents
   (b) Odorants.

6. Other Additives

   (a) Curing Agents
   (b) Clarifying Agents
   (c) Chain Extenders
   (d) Accelerators
   (e) Slip or Antislip Agents
   (f) Anti-Plate-Out Additives
   (g) Antiplasticisers.

7. Additives from Natural Sources.

Considering the scope of this publication, the following text will only address a number of selected additives.

1.2.1 *Lubricants*

Lubricants are used to enhance the flow properties of polymers. They reduce friction, which significantly aids their processing. Pritchard [14] states that certain additives that are used as lubricants frequently also have other functions. Some of them help to stabilise PVC or act as slip or antiblock agents. Some even improve antistatic behaviour by increasing the surface lubricity. Several PVC aliphatic esters such as adipates, palmitates, and sebacates are both lubricants and plasticisers.

1.2.2 Plasticisers

Polymer materials are required to have high flexibility and pliability within certain applications. Plasticisers are used to achieve these properties. They typically include fluids with high boiling points [15], which after having been mixed with a polymer reduce the number of internal and intermolecular bonds. Generally, this can be achieved by the introduction of atomic groups or molecules into the polymer that increases the distance between the individual chains, separate polar groups, etc. [16]. By doing so, the distance between the sections of the polymer chains and the whole molecules and their mutual mobility is increased [2]. For this reason, plasticisers are sometimes described as internal lubricants.

The vast majority of plasticisers are used to plasticise PVC. The reason is that only a small number of polymers meet both of the following criteria at the same time [14]:

1. Compatibility with the plasticiser
2. A medium, but not too high, degree of crystallinity to allow it to retain elasticity after it is mixed with a plasticiser.

The normal concentration of plasticisers is 20–40%, but some systems have 50–60% plastification [17]. Substances used as plasticisers with polymers include, for example, camphor, triphenyl phosphate, dibutyl phthalate, tricresyl phosphate, or dimethyl phthalate [15].

1.2.3 Colourants

As the name itself suggests, colourants change the colour of polymers. Subramanian [13] classifies them into dyes and pigments. While dyes are soluble organic substances, pigments are insoluble and may be either organic or inorganic. Colourants are commonly added to materials in amounts equal to 1–4% of their weight. Just as with other additives, they may also affect particular properties of the resultant material. For instance, titanium dioxide, which is often used as a white colourant, could make the plastic harder and reduce its flexibility [18].

1.2.4 Fillers

Fillers are additives that are added into polymers in large quantities to reduce costs by replacing an amount of the plastic with a cheaper material. But [14] warns that the cost saving may be very low or even zero, since the weight of the product may increase, or that specific processes may be necessary to mix the filler with the polymer, and that it is sometimes difficult to modify the surface of the filler particles. In many cases, the

Table 1.2 Examples of flammable and non-flammable fillers

Flammable fillers	Non-flammable fillers
Sawdust	Talc
Wood flour	Glass fibres
Textile fibres	Kaolin
Paper pulp	Asbestos
Linen fibres	Sand
Starch	Mica
Soy protein	Barium sulphate
	Calcium carbonate
	Magnesium silicate

fillers used often fulfil an additional function and modify the properties of the polymer material in the desired direction. For example, by adding 30% short glass fibres by weight to Nylon 6 the creep resistance is improved and stiffness increases by 300% [19]. Both inorganic (such as talc and glass fibres) and organic (such as sawdust and wood flour) additives are utilised. Considering they are used in large volumes, they could have a substantial impact on thermal degradation and combustion properties of the resulting material. Examples of flammable and non-flammable fillers are listed in Table 1.2.

In order for a material to be usable as filler, it must fulfil the following 9 conditions [15]:

1. Compatibility and ease of mixing with resins and other additives
2. Ease of moulding
3. Absence of abrasive properties
4. No chemical reaction with moulds
5. Good electrical characteristics
6. High heat resistance
7. Low moisture absorption
8. Low cost
9. Abundant supply.

1.2.5 Stabilisers

Polymer materials tend to be sensitive to their surroundings. Higher temperatures, UV radiation, oxygen, or even weather conditions can lead to their degradation. They negatively affect plastics, and they lose their required properties and may be subject to colour changes, embrittlement, or cracking. Additives used to protect them from these effects are called stabilisers.

Due to the existence of chromoform groups in most polymers, photochemical reactions triggered by exposure to light lead to their degradation. This exposure may cause a rearrangement of the polymer molecule or even break internal bonds, and not only where the light energy was absorbed. Radiation in the range of wavelengths between 290 and 400 nm, which represents approximately 5% of the total solar radiation that reaches the surface of the Earth, has the most significant effect [16]. Plastics may be stabilised against the effects of UV radiation using substances that reflect it (such as titanium dioxide) or absorb it (such as soot or zinc oxide). For transparent plastics, organic substances are used, which do not colour the polymer and are able to effectively absorb UV radiation [2].

When oxidising agents (most commonly oxygen) affect polymers, they start to oxidise. Under normal circumstances, the impact of oxygen is unremarkable, and oxidation occurs very slowly. However, when several conditions are combined, such as higher temperatures or the presence of UV radiation the rate of oxidation can be accelerated. Even a small degree of oxidation can trigger a sudden decrease in molecular weight and along with it a change in properties of the polymer [16]. According to Pritchard [14], antioxidants are only used in relatively small amounts ranging from 0.05 to 0.25%, with their molar concentration often being more important than the dose by weight. Ram [17] suggests that a usable concentration is between 0.01 and 1%. Phosphoric acid esters, phenols, aromatic amines, and other organic compounds are used as antioxidants [2].

Heat stabilisers are additives that enable plastics to be used for higher temperature applications. If the plastic decomposed under these conditions, not only might there be some deterioration in their properties, but they could also generate toxic by-products. Heat stabilisers may be divided into two groups [1]:

1. Stabilisers based on the salts of inorganic and organic acids, which contain cations of lead, strontium, zinc, magnesium, lithium, calcium, sodium, and so-called organometallic stabilisers
2. Organic stabilisers, which include, for example, epoxy compounds, organic phosphides, stabilisers based on urea and its derivatives, and esters of β-aminocrotonic acid.

1.2.6 Blowing Agents

Blowing agents are used in the production of foam materials based on plastics, such polystyrene foam. They release gaseous substances in the temperature range between the melting point and the decomposition temperature of a polymer. Once the polymer composite heats up, the melted polymer starts to foam. After the material cools down, it hardens, creating a porous material with an open and closed cell structure. The release of gas is either based on a physical phenomenon or is due to a chemical reaction. In chemical foaming, a specific mass of about 0.5 may be obtained, compared to as little as 0.03 for physical foaming [17].

The most commonly used physical blowing agents used to include chlorofluorocarbons, which are no longer used due to their negative environmental effects. They were replaced by, for instance, pentane or cyclopentane [14]. Physical blowing

Table 1.3 Examples of chemical blowing agents [20]

Blowing agents	Type	Decomposition temperature [°C]	Evolved gases
p-Toluenesulphonylhydrazide	Exo	110–120	N_2, H_2O
Sodium bicarbonate	Endo	120–150	CO_2, H_2O
4,4-Oxybis(benzenesulphonyl-hydrazide)	Exo	150–160	N_2, H_2O
Dinitrosopentamethylenetetramine	Exo	195	N_2, NH_3, HCHO
Citric acid derivatives	Endo	200–220	CO_2, H_2O
Azodicarbonamide	Exo	200–230	N_2, CO, NH_3, O_2
p-Toluenesulphonylsemicarbazide	Exo	215–235	N_2, CO_2
5-Phenyltetrazole	Endo	240–250	N_2
Polyphenylene sulphoxide	Exo	300–340	SO_2, CO, CO_2

agents also include gases that are introduced into the melted polymer using pressure as part of the production process. Carbon dioxide, nitrogen, and short-chain aliphatic hydrocarbons tend to be used this way [20].

The use of chemical blowing agents (Table 1.3) results in the decomposition of the original molecule to provide one or more gases for polymer expansion, and one or more solid residues that remain in the foamed polymer [20]. The first chemical blowing agents used were ammonium carbonate and sodium bicarbonate. However, other types of chemical blowing agents later also became used, especially ammonium and sodium salts. These blowing agents are cheap, however, they are difficult to disperse in polymers and start to freely decompose if they are stored for a longer period of time [1]. More than half of all the commercially available chemical blowing agents are used with PVC, but they can also be employed with polyolefins, polystyrene, PET, and rubber [14].

1.2.7 Fire Retardants

Most polymers are flammable as their composition often includes a large proportion of flammable elements. To ensure they are useable in practice, it is often necessary to reduce their flammability through the use of fire retardants. There are various mechanisms of fire retardation, and they are often combined. Based on their mechanism of retardation, they are divided into 4 groups. Their mechanism of retardation may be due to [21]:

1. Gas phase retardation
2. Cooling
3. Decreasing oxygen concentration
4. Solid state retardation.

1.2 Additives Added to Polymers

In gas phase retardation, the retardant's molecule breaks down into free radicals. These react with the radicals in the flame, which are responsible for the chain reaction of combustion. This causes a slowdown and subsequent termination of the reactions in the flame. In many cases, the radicals within the fire retardants can restore themselves, thus decreasing the amount of retardant necessary. This group typically includes halogen derivatives of hydrocarbons from which halogen radicals are released by heat. There are other fire retardant additives, among them, for example, antimony trioxide.

Retardants that cool the combustion area are substances that decompose endothermically. As they decompose, they use up the heat that would otherwise heat up the polymer. These often include inorganic hydrates that through their endothermic reaction release water in the form of steam. This type of fire retardant includes, for example, aluminium hydroxide or magnesium hydroxide.

Certain types of retardants decompose to non-flammable gases under heat. These gases are then released, which decreases the oxygen concentration in the combustion zone. Examples of this type of retardant are the aforementioned aluminium hydroxide or magnesium hydroxide, which release water vapour during decomposition.

Solid phase retardation is based on the formation of a barrier between the area where combustion is taking place and the polymer. In this case, it is the degraded surface of the polymer that serves as a barrier. It can be formed in one of two ways. The first is through carbonisation. In the process of carbonisation, the retardant initiates cross-linking at increased temperatures. By doing so, it triggers a loss of hydrogen atoms in the polymer but the carbon atoms are connected to more stable structures through chemical bonds. These reactions are catalysed by the acidic conditions produced by certain types of cations. The carbonised layer can also be produced by non-flammable retardant residues, or even through highly thermally stable substances, which float to the surface of the molten polymer layer and thus separate it from the combustion zone. Commonly used retardants include silicone and phosphorous. The second way to create a barrier is to form a non-flammable foam through a process called intumescence. In the first phase, the polymer melts and the fire retardant decomposes releasing acid. This is followed by esterification—the released acid reacts with components rich in carbon. The resulting esters mix with the molten polymer and decompose, and a carbon-inorganic residue is produced. Simultaneously, gaseous products are released, leading to the foaming of the carbonising material. Finally, the material solidifies in the form of a multi-cellular foam. This process resembles the foaming of polymers by chemical blowing agents; however, the latter occurs in fires and results in the production of a material rich in carbon and inorganic components. The layer formed in this process has a low coefficient of heat transfer; thus, it insulates the non-degraded plastic layer from the heat source. At the same time, it prevents flammable pyrolysis products from entering the combustion zone. Retardants of this type must be optimised for a specific polymer. In the past, mixtures of alcohol, ammonium components, and phosphorous were used. However, they posed the problem of the incompatibility of the individual components and polymer. Recently, however, there has been an increase in the effort to develop retardants that contain all the components within a single molecule.

In addition to the use of retardant additives in polymers, a technology which involves their addition to reaction mixtures in the synthesis of polymers has also been used, after which they remain chemically bound to the macromolecular chain. An alternative approach is to apply a protective layer of retardant on the surface of a plastic product [2].

1.3 The Impact of Increased Temperature on Polymers

If a polymer is exposed to increased temperatures, the kinetic energy of the macromolecules in certain polymers increases. Consequently, they soften and begin to melt. Alternatively, no melting occurs but the polymer chain degrades and decomposes. Macromolecules become brittle and release volatile, often flammable products.

Softening and melting of plastics are characterised by the glass transition temperature and the melting point (Table 1.4). The glass transition temperature is described as the temperature at which 30–50 carbon chains start to move. At the glass transition temperature, the amorphous regions experience transition from rigid state to more flexible state making the temperature at the border of the solid state to rubbery state. The melting point is the critical temperature above which the crystalline regions in a semi-crystalline plastic are able to flow [22].

From the perspective of thermal degradation of polymers, their molecular structure plays a pivotal role. They may have linear, branched, or cross-linked structures. Linear molecules are long fibres formed by monomer units connected one after another, without any side chains. Branched molecules have areas where another chain is bound. The molecule does not have the shape of a single fibre, but it consists of more fibres connected on one side to a different fibre of the same molecule. Polymer molecules with a cross-linked structure contain mutually interconnected chains that create a network. As a rule, this form is significantly more stable than the linear or branched structures, and this is reflected in the thermal resistance. Furthermore, the presence of aromatic nuclei in the molecule also leads to higher thermal resistance.

Ducháček [1] states that a polymer with a significantly high-molecular weight has a boiling point that is higher than the decomposition temperature. Hence, polymers may only exist in a liquid or solid form.

Due to changes in the chemical and physical properties when heated, the suitability of polymers for specific purposes is restricted as the surrounding temperature must be lower than the threshold temperature. Threshold temperatures for the long-term use of polymers are indicated in Table 1.5.

Thermal degradation of polymers can follow three major pathways [25]:

1. Side-group elimination—first, the side groups are eliminated and subsequently the unstable polyene macromolecule undergoes further degradation, including the formation of aromatic molecules, scission into smaller fragments, or the formation of char

1.3 The Impact of Increased Temperature on Polymers

Table 1.4 Glass transmission temperatures and melting points of various polymers

Polymer	Glass transition temperature [°C]	Melting point [°C]	Source
Polylactic acid	58.5	150.5	[23]
Polyethylene (low density)	−110	115	[24]
Polytetrafluoroethylene	−97	327	[24]
Polyethylene (high density)	−90	137	[24]
Polypropylene	−18	175	[24]
Poly(ethylene terephthalate)	69	265	[24]
Poly(vinyl chloride)	87	212	[24]
Polystyrene	100	240	[24]
Polycarbonate	150	265	[24]
Natural rubber	−73	36	[15]
Polyacrylonitrile	97	341	[15]
Polybutadiene (cis)	−102	6	[15]
Polybutadiene (trans)	−58	100	[15]
Polyethylene (high density)	−125	146	[15]
Polyethylene terephthalate	69	264	[15]
Polyisobutylene	−73	44	[15]
Polymethyl methacrylate (isotactic)	38	160	[15]
Polypropylene (isotactic)	−8	208	[15]
Polystyrene	100	250	[15]
Polyvinylchloride	81	310	[15]

2. Random scission—the formation of a free radical at some point on the polymer backbone, producing small repeating series of oligomers usually differing in chain length
3. Depolymerisation—the reverse mechanism to polymerisation, in which the formation of a free radical on the backbone causes the polymer to undergo scission to form unsaturated small molecules and propagate to the free radical on the polymer backbone.

In case of depolymerisation, the polymer chain breaks down without any disruption to the individual monomers. The latter, however, occurs in polymers whose macromolecule does not contain groups capable of chemically reacting at depolymerisation temperatures. In this case, the release of monomers may occur [1].

Table 1.5 Threshold temperature for the long-term use of selected polymers [1]

Polymer	Threshold temperature for long-term use [°C]
Polyvinyl acetate	35
Polyvinylchloride	60
Pieces of natural rubber	70
Polyethylene	75
Polystyrene	80
Pieces of butadiene rubber	80
Polyamide	80–120
Amino plastics	80–140
Polybutylene	90
Polyoxymethylene	90
Cellulose and its derivates	100
Polypropylene	100
Polyphenylene oxide	100
Polyisobutylene	100
Pieces of butyl rubber	100–140
Epoxy resins	100–150
Phenoplasts	100–150
Polymethylmethacrylate	110
Polyvinyl formal	120
Polycarbonate	130
Polyvinyl butyral	130
Polyvinylfluoroethylene	150
Pieces of silicone rubber	180–200
Polytetrafluoroethylene	250

1.4 The Flammability of Polymers

As mentioned above, polymers often contain a relatively large amount of flammable elements, which results in the flammability of the resulting polymers. But if, for example, halogens are introduced into a molecule, its flammability decreases. In addition to the composition of the polymer, its flammability is also affected by the structure of the molecules and the additives, fillers, or physical properties of the final product. One of the properties that characterises materials from the perspective of their combustibility is the limiting oxygen index. This is the lowest possible concentration of oxygen that still allows burning to take place under the given conditions. This provides us with a simple method to compare various materials. Its values for selected polymers are shown in Table 1.6.

1.4 The Flammability of Polymers

Table 1.6 Values for the oxygen index of various polymer materials

Polymer	Chemical formula	Limiting oxygen index	Source
Polyacrylonitrile	$(C_3H_3N)_n$	16.9	[26]
Polyethylene	$(C_2H_4)_n$	18.4	[26]
Polystyrene	$(C_8H_8)_n$	17.8–18.1	[16]
Polyethylene terephthalate	$(C_{10}H_8O_4)_n$	22.7	[26]
Polyvinyl chloride	$(C_2H_3Cl)_n$	45–47	[16]
Polytetrafluoroethylene	$(C_2F_4)_n$	95	[26]
Cellulose	$(C_6H_{10}O_5)_n$	18.0	[27]
Polypropylene	$(C_3H_6)_n$	22	[28]
Ethylene–vinyl acetate (18% vinyl acetate content)	$(C_2H_4)_n + (C_4H_6O_2)_m$	17.5	[29]
60% Ethylene–vinyl acetate (18% vinyl acetate content) 40% Magnesium hydroxide	$(C_2H_4)_n + (C_4H_6O_2)_m + Mg(OH)_2$	22	[29]
50% Ethylene–vinyl acetate (18% vinyl acetate content) 50% Magnesium hydroxide	$(C_2H_4)_n + (C_4H_6O_2)_m + Mg(OH)_2$	24	[29]
40% Ethylene–vinyl acetate (18% vinyl acetate content) 60% Magnesium hydroxide	$(C_2H_4)_n + (C_4H_6O_2)_m + Mg(OH)_2$	42.5	[29]
Linear low density polyethylene	$(C_2H_4)_n$	17.5	[30]
50% Linear low density polyethylene 50% Magnesium hydroxide	$(C_2H_4)_n + Mg(OH)_2$	22.5	[30]
40% Linear low density polyethylene 60% Magnesium hydroxide	$(C_2H_4)_n + Mg(OH)_2$	25.5	[30]
Polyvinyl fluoride	$(C_2H_3F)_n$	22.6	[16]
Polychlorotrifluoroethylene	$(C_2ClF_3)_n$	95	[16]
Polyphenylene oxide	$(C_8H_8O)_n$	28–30.5	[16]
Polylactic acid	$(C_3H_4O_2)_n$	19	[31]

The combustion of polymers can be explained in a simple way in Fig. 1.2. A polymer exposed to a heat flux degrades and releases gaseous products of degradation. These diffuse into and are mixed with the surrounding atmosphere (almost exclusively with the air). After a sufficient temperature and suitable ratio between flammable gases and an oxidising agent are reached, initiation of flame combustion takes place. The ignitor could be an existing flame. The oxidation of the flammable gases releases heat that is partially transferred to the environment and partially heats the polymer. Provided that the speed of heat release is sufficiently high, combustion can be maintained and the flame may even spread over the surface of the polymer.

However, when the polymer is exposed to heat, some of it does not turn into gaseous substances. They may be either liquid or solid. The liquid products of combustion, which are not released as vapour due to the heat, may produce run-off into the surrounding area. Solid material creates a layer on the surface of the flammable substance called the char layer. This is mostly composed of carbon atoms that are mutually bound by the more solid chemical bonds. The char layer does not release any further volatile combustible matter, but if heated to a high temperature and in contact with an oxidising agent, it triggers smouldering. This process is significantly slower than flame combustion.

Moreover, a fire plume is generated during combustion. It may be defined as the motion generated by a source of buoyancy which exists by virtue of combustion and may incorporate an external source of momentum. The buoyancy source may be due to glowing or flaming combustion of a solid or liquid, with no external source of momentum, or due to gaseous, liquid, spray, or aerosol discharge from an opening at various mixes of mass flow and momentum [32].

Hilado [33] describes the process of polymer burning using three scales (Table 1.7). In the case of microscale combustion, he mostly observes the behaviour of a polymer material on a molecular level. On the macroscale, he described the behaviour

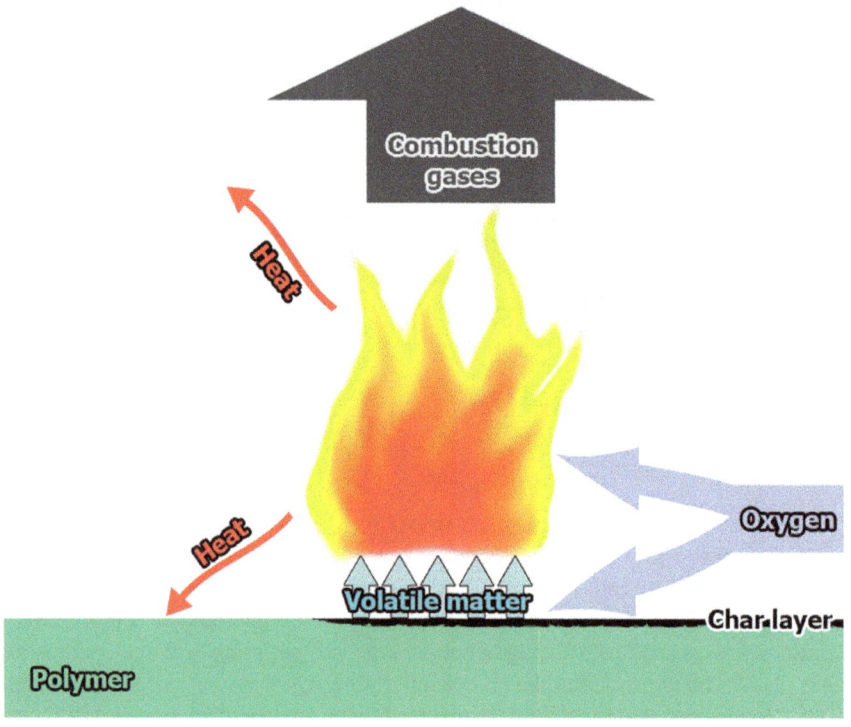

Fig. 1.2 Schematic illustration of the process of combustion of polymer materials

1.4 The Flammability of Polymers

of a material with regard to its weight and the mass scale is then used to describe the behaviour of a whole system, such as a room or a building.

Table 1.7 Description of combustion at various scales (according to Hilado [33])

Scale	Stage	Description
Micro	Heating	The polymer is heated by an external heat source. There are only minor changes to physical properties
	Transition	In a narrow temperature range, the polymer changes into a viscous, gum-like state. There are rapid changes in its mechanical and some thermal properties
	Degradation	Chemical bonds with low thermal resistance break down, with a large portion of the macromolecule remaining stable, this may lead to a change in colour
	Decomposition	The majority of bonds break down. Depending on the type of polymer, various phenomena occur ranging from the decomposition of the original structure into monomers to the creation of new structures. There is only a small weight loss
	Oxidation	Fragments of the polymer react quickly with oxygen. A flame and possibly glowing of the carbon residues occur
Macro	Heating	An increase in the temperature of a polymer by an external heat source
	Decomposition	The polymer reaches its decomposition temperature and starts to release products such as combustible gases, non-combustible gases, liquids, solids, and entrained solid particles
	Ignition	The ignition of flammable gases occurs in the presence of an oxidising agent (oxygen in the air)
	Combustion	Combustion releases part of the heat of combustion, this increases the temperature of the gaseous products and non-flammable gases, and the heat transfer increases. For small sections of the polymer, this state represents fully developed burning
	Propagation	A proportion of the heat of combustion brings adjacent unit mass to the combustion stage. This leads to the propagation of burning
Mass	Initial fire	In the initial stages of burning, an initiating source affects a flammable substance. Plastics are almost never ignitors, but they may be the first substance ignited by an ignitor
	Fire build-up	The heat produced by burning accumulates in the system, and the temperature of the materials rises through conduction, convection, and radiation. However, the fire itself spreads slowly
	Flashover	The point at which most of the flammable materials in the system reach ignition temperature and simultaneously begin to burn
	Fully developed fire	Practically, all flammable materials contribute to the fire
	Fire propagation	The heat released from the fire is sufficient to spread burning to surrounding systems

Table 1.8 Yields of carbon monoxide and carbon dioxide when a material is exposed to an external heat flux of 25 kW m^{-2} (Shi and Chew [130])

Polymer	Non-flaming combustion		Flaming combustion	
	Yield of carbon monoxide [g g^{-1}]	Yield of carbon dioxide [g g^{-1}]	Yield of carbon monoxide [g g^{-1}]	Yield of carbon dioxide [g g^{-1}]
HDPE	0.12	0.052	0.021	1.99
PP	0.02	0.037	0.03	2.16
PMMA	0.007	0.04	0.02	2.00
ABS	0.0028	0.016	0.086	2.71
PET	0.09	0.09	No flame observed	No flame observed
PC	0.19	0.74	No flame observed	No flame observed

Unlike thermal degradation, where polymer scission can occur randomly and/or at the chain end, oxidative degradation is characterised by random scission in the polymer backbone [25]. If we carry out a thermogravimetric analysis in the air, an increase in the weight of the polymer can be observed at the beginning of the measurement. This is probably triggered by a small degree of polymer oxidation before thermal degradation [34].

The products of burning generated by polymers mostly include carbon oxides, water, and a mixture of organic substances that originate from polymer chains and often contain oxygen groups. An overview of the yield of carbon oxides in flaming and non-flaming combustion of various polymers is provided in Table 1.8. In the case of flaming combustion, the amount of carbon dioxide released is significantly higher, and the release of carbon monoxide is lower in comparison with non-flaming combustion.

1.5 The Thermal Degradation of Selected Polymers

1.5.1 Polypropylene (PP)

The polymerisation of propylene produces polypropylene (Fig. 1.3). Radical and cationic polymerisation will only produce a low-molecule product consisting of branched atactic molecules. However, when using certain catalysers, it is possible to achieve a high-molecular polypropylene with a regular structure, high melting point and good mechanical properties [16]. The high degree of crystallinity (60–75%) makes the material non-transparent. Its melting point in a clear state is 176 °C [1].

The thermal degradation of PP essentially occurs through two reaction steps [35, 36]:

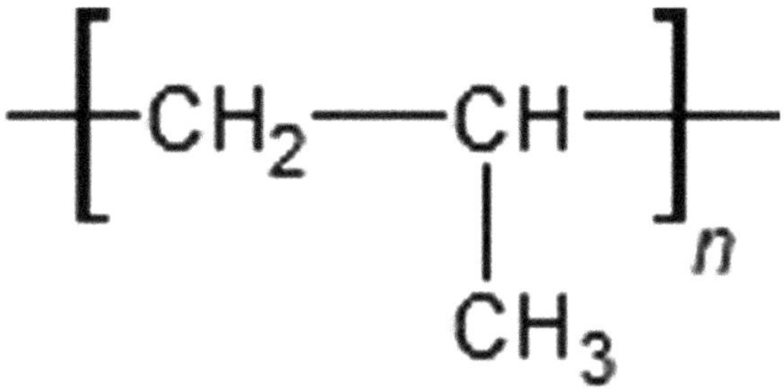

Fig. 1.3 Chemical formula of polypropylene

$$\left[-CH_2-CH- \atop {| \atop CH_3} \right]_n$$

1. Depolymerisation, also called unzipping, during the production of propylene
2. Hydrogen transfer from the tertiary carbon atom along the polymer chain to the radical site and β-scission.

Canetti et al. [133] state that polypropylene volatilises in a single step, from 280 to 500 °C, without char formation when exposed to a thermal load in a nitrogen atmosphere. Decomposition in air also only occurs in a single step, but in the temperature range of 200–400 °C. Similar data has been published by Bertini et al. [134], who determined the beginning of volatilisation in nitrogen at around 270 °C and the temperature range of decomposition in air at 200–420 °C. Piloted ignition of polypropylene due to radiative heating has been observed at a surface temperature of 610 K.

More than 90% of the carbon in polypropylene is released during pyrolysis in the form of volatile organic compounds such as dienes, alkanes, and alkenes [37].

During the combustion of polypropylene in air at temperatures of 200–600 °C, a wide range of different products are formed, including aldehydes, ketones, ethers, organic acids, alcohols, aromatic hydrocarbons, aliphatic hydrocarbons, carbon dioxide, carbon monoxide, and water. The amount of the different products depends on various factors such as temperature, combustion time, and the air to fuel ratio. The major oxygenated hydrocarbons that have been identified are acetone, acetaldehyde, formaldehyde, acetic acid, and methanol. The major aromatic hydrocarbons that have been detected during combustion are benzene, toluene, methylethylbenzene, xylene, and styrene [38].

1.5.2 Polyethylene (PE)

Although from a chemical perspective, polyethylene is a homopolymer of ethylene (Fig. 1.4), in practice this term is also used for its copolymers with a small amount of added co-monomer. Their properties are highly dependent on molecular weight, the spatial organisation of the mere in the macromolecular chain and the degree of crystallinity [1]. Depending on the production conditions, it is possible to produce polyethylene with varying densities. Thus, it is accordingly classified into different groups as specified in Table 1.9. This classification is not globally unified; for this reason, the indicated values correspond with the [39] and ASTN 1248 standards.

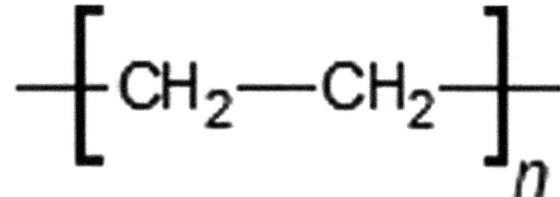

Fig. 1.4 Chemical formula of polyethylene

Table 1.9 Classification of types of polyethylene based on density [39, 41]

Class	ISO 17855		ASTM 1248	
	Abbreviation	Range of density [kg m^{-3}]	Abbreviation	Range of density [kg m^{-3}]
Ultra-low density polyethylene	–	–	ULDPE	890–905
Very low density polyethylene	PE-VLD	≤911	VLDPE	905–915
Low density polyethylene	PE-LD	911–925	LDPE	915–935
Linear low density polyethylene	PE-LLD	911–925	LLDPE	915–935
Medium density polyethylene	PE-MD	925–940	MDPE	926–940
High density polyethylene	PE-HD	>940	HDPE	940–970

In addition to the specified classes, it is possible to come across other names, such as low-medium density polyethylene (LMDPE) [40], ultra-high molecular weight polyethylene (UHMWPE) [41], or cross-linked polyethylene (PEX) [42].

The production of polyethylene is primarily focussed on two types: low density polyethylene and high density polyethylene. PE-LD is produced by polymerisation under high pressure, and its molecules are branched. On the other hand, PE-HD is formed under low pressure and its structure is linear [2].

Low density polyethylene melts at 110–125 °C and is a partially, 50–60%, crystalline solid. While there is practically no solvent that dissolves it at room temperature, it is soluble in many solvents at temperatures above 100 °C. Some of the solvents that can be used to dissolve polyethylene at higher temperatures are carbon tetrachloride, toluene, decaline, trichloroethylene, and xylene. High density linear polyethylenes are highly crystalline polymers that contain less than one side chain per 200 carbon atoms in the main chain [15].

In an inert atmosphere, polyethylene begins to cross-link at 475 K and to decompose (reductions in molecular weight) at 565 K although extensive weight loss is not observed below 645 K. Piloted ignition of polyethylene due to radiative heating has been observed at a surface temperature of 640 K [43].

The thermal degradation of polyethylene follows two different kinds of pathways. These are random and chain-end scissions which include b-scission on the chain-end and radical transfer scission [44]. Zong et al. [45] state that although random scission is a primary degradation pathway of polyethylene, it can also result in polymer chain branching. Both scission and branching occur simultaneously and give rise to a single mass loss step.

In nitrogen, high density polyethylene only volatilises at temperatures higher than 410 °C and total degradation is only reached at about 500 °C [46]. LDPE pyrolysis occurs at lower temperatures than in the case of HDPE [47, 48] (Kremer et al. [131]). Dubdud and Al-Yaari explain that the higher decomposition temperature of LDPE is caused by the higher degree of branching. They suggest that LDPE pyrolysis occurs in the temperature range of 172–512 °C and in the case of HDPE between 171 and 517 K.

Because unzipping is insignificant in polyethylene, the degradation process in air primarily occurs via reactions with oxygen [34].

Polyethylene thermally degrades without any residue to a large quantity of paraffinic and olefinic compounds. The analysis of pyrolysis products with GC/MS reveals high amounts of linear-alkanes and n-alkenes. The quantity of dienes is low. Neither branched, aromatic or cyclic compounds, nor Diels–Alder products from butadiene have been detected [49]. Hodgkin et al. [50] explain that the pyrolysis products of polyethylene include a range of saturated and unsaturated hydrocarbons from C2 to C23 whose ratio of production does not vary greatly with changes in conditions. The oxidative degradation products, mainly including acetone, acetaldehyde, acetic acid, and a small amount of acrolein, considerably vary in relative yield. When polyethylene was burnt in a horizontal furnace, more than 230 degradation products were discovered. Higher concentrations of oxygen led to a decrease in the yield of hydrocarbons, while the amount of semi-volatile compounds including oxygen and carbon oxides increased. With increased temperatures both cracking and pyrosynthesis reactions were enhanced, which lead to an increase in the yields of methane, ethane, ethylene, benzene, or polyaromatic hydrocarbons, whereas the production of other linear hydrocarbons, oxygenated compounds, and carbon oxides fell [51]. Ballice et al. [52] found that the maximum volatile product formation temperature is 425 °C for PE-LD and 430 °C for PE-HD. PE-HD gave a higher yield of gaseous products and is more difficult to degrade than PE-LD.

1.5.3 Polyvinylchloride (PVC)

Polyvinylchloride is the product of the polymerisation of vinylchloride, and due to its relatively low cost and wide range of applications, it is one of the most commonly used synthetic polymers. Pure PVC is a white, brittle solid and in practice is used in two basic forms: rigid and flexible [53]. Its chemical formula is shown in Fig. 1.5.

Polyvinyl chloride does not have a completely regular structure; therefore, it is called a partially syndiotactic material. It has low crystallinity. The polymer molecule

Fig. 1.5 Chemical formula of polyvinylchloride

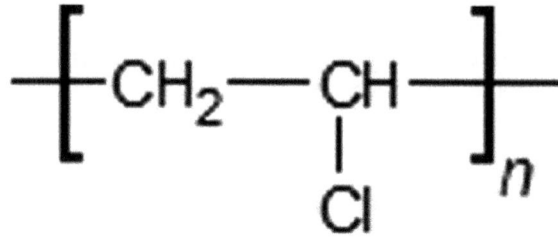

is complicated by the possibility of having either a linear structure or a purely branched structure. It is insoluble in water, hydrocarbons, vinyl chloride, and alcohols. It is unaffected by acids and alkalis at temperatures up to 20 °C. It is soluble in ketones, chlorinated hydrocarbons, and a mixture of acetone and carbon disulphide [15].

PVC cannot be processed on its own due to its very low thermal stability and high melt viscosity. Therefore, it is necessary to combine a number of suitable additives with the polymer to give a wide and varied range of properties to satisfy many different end-use applications [54].

In thermogravimetric measurements, PVC exhibits two degrees of degradation both in air and in an inert atmosphere. The first ranges from 270 to 360 °C, where atmospheric oxygen has no effect, and the second, in particular, between 400 and 500 °C, albeit continuing up to 800 °C. The course of the second reaction in air seems to involve more individual steps and leads to more extensive weight loss [55].

Based on measurements taken in vacuum at temperatures up to 500 °C, McNeil et al. [56] classify the thermal degradation of polyvinylchloride as follows:

- The first stage (from 200 to 360 °C): Mainly, HCl and benzene and very little alkyl aromatic or condensed ring aromatic hydrocarbons are formed. It was evaluated that 15% of the polyene generates benzene, the main part accumulating in the polymer and being active in intermolecular and intramolecular condensation reactions by which cyclohexene and cyclohexadiene rings embedded in an aliphatic matrix are formed;
- The second stage (from 360 to 500 °C): Alkyl aromatic and condensed ring aromatic hydrocarbons and very little hydrogen chloride and benzene are formed. The polymeric network formed by polyene condensation breaks down in the process of aromatisation of the above C_6 rings.

A comparison of the products of thermal degradation of PVC at different temperatures in an inert atmosphere and in air as determined by [57] is shown in Table 1.10. The main difference is the presence of carbon oxides, which are not produced in an inert atmosphere due to the absence of oxygen. Yet, in air, they comprise a substantial part of the released gases, especially at higher temperatures. With increasing temperature, the proportion of carbon dioxide also increases.

Owing to the chlorine atom in its molecule, PVC has a low degree of flammability. When the surface of rigid PVC is exposed to a flame, a charred layer is produced. It

1.5 The Thermal Degradation of Selected Polymers

Table 1.10 Products of the thermal degradation of PVC [57]

Temperature [°C]	350	600	850
Thermal degradation in helium			
Products	Hydrogen chloride	Hydrogen chloride	Hydrogen chloride
	Benzene	Benzene	Benzene
	Toluene	Methane	Methane
		Ethane	Ethene
		Toluene	Toluene
		Ethene	Hydrogen
		Hydrogen	Ethane
Thermal degradation in air			
Products	Hydrogen chloride	Hydrogen chloride	Hydrogen chloride
	Benzene	Carbon dioxide	Carbon dioxide
	Carbon dioxide	Carbon monoxide	Carbon monoxide
	Carbon monoxide	Benzene	Methane
		Methane	Benzene
		Ethene	Ethene
		Toluene	Toluene
		Hydrogen	Hydrogen
		Ethane	Ethane

insulates the material below and excludes the oxygen necessary for combustion. This restricts the burning zone. The hydrogen chloride emitted also acts as a combustion inhibitor. Its peak rate of heat release is low in comparison with other materials and so does not release enough heat to support its own combustion. When the flame source is removed or extinguished, the PVC ceases to burn [54].

1.5.4 Polyurethanes (PUR)

Polyurethanes may be defined as polymers created by the reaction of multifunctional isocyanates with polyalcohols. Most polyurethanes are based on aromatic polyisocyanates, which are more reactive and cheaper than aliphatic compounds [16]. An example of the chemical formula of polyurethane is shown in Fig. 1.6. As explained by Szychers [58], the term polyurethanes are quite confusing because polyurethanes are not derived from the polymerisation of a methane monomer, nor are they polymers that primarily contain urethane groups. The polyurethanes include those polymers that contain a plurality of urethane groups in the molecular backbone, regardless of the chemical composition of the rest of the chain. Thus, a typical polyurethane may contain, in addition to the urethane linkages, aliphatic and aromatic hydrocarbons, esters, ethers, amides, urea, and isocyanurate groups [58].

From the perspective of their chemical resistance, polyurethanes resist acids and bases, carbons, polar organic solvents, fats, and oils. They dissolve in hydrofluoric acid and partially in halogen derivatives of hydrocarbons [2]. Polyurethanes may

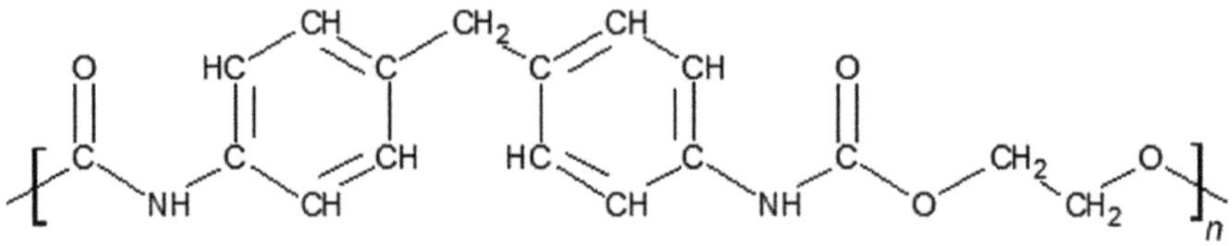

Fig. 1.6 An example of the chemical formula of polyurethane (processed according to Rahman et al. [59])

occur as crystalline solids, segmented solids, amorphous glasses, or viscoelastic solids. From the perspective of their mechanical properties, they are non-ideal solids [58].

The mechanism of thermal decomposition of polyurethanes depends on the structure of each polymer [60]. It decomposes through random-chain scission, chain-end scission (unzipping), and cross-linking. The most common mechanisms are random-chain scission and cross-linking. In the case of polymers with a cross-linked structure, their thermal degradation begins with the degradation of the side chains. In general, thermal degradation of polyurethanes occurs in the following steps [61]:

1. Release of trapped volatile materials
2. Scission and depolymerisation resulting in weight loss and degradation of mechanical properties
3. Complete thermal breakdown of the chains.

Polyurethanes degrade at low temperatures (200–300 °C) with the formation of a nitrogen-free residue and yellow smoke that contains nitrogen. With increased temperature, the residue further decomposes to smaller compounds, and the yellow smoke yields nitrogen-containing products like hydrogen cyanide and acetonitrile [25].

During thermal degradation, the rate of decomposition tends to be slowed by air. The production of hydrogen cyanide and carbon monoxide increases as the pyrolysis temperature increases. Other toxic products formed include nitrogen oxides, nitriles, and isocyanates. A major breakdown mechanism in urethanes is the scission of the polyol–isocyanate bond formed during polymerisation [43]. Chattopadhyay and Webster [61] summarised the products of the thermal degradation of polyurethanes as a mixture of simple hydrocarbons, carbon monoxide, carbon dioxide, hydrogen cyanide, methanol, acetonitrile, acrylonitrile, propionitrile, pyrrole, pyridine, aniline, benzonitrile, quinoline and phenyl isocyanate, and a complex char.

1.5.5 Polystyrene (PS)

Polystyrene is the polymer of styrene (Fig. 1.7). In its basic state, it is a transparent, hard, and quite rigid plastic. It is particularly well known for its use in the production

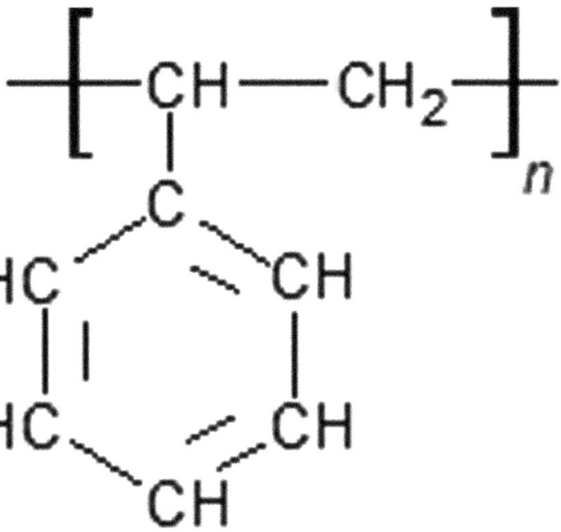

Fig. 1.7 Chemical formula of polystyrene

of CD cases, but it is also used in other areas, including the production of plastic cups. However, most people know it in its foam form, which is produced through the addition of a blowing agent, usually pentane, cyclopentane, or carbon dioxide. In this state, it is available as expanded polystyrene (EPS) and extruded polystyrene (XPS).

Polystyrene degradation in nitrogen starts at about 360 °C, and the polymer is completely decomposed at 450 °C [46].

Thermal degradation of polystyrene in air occurs at a single stage between 250 and 400 °C. Similarly, as is the case with many other polymers, degradation occurs in the presence of oxygen at lower temperatures than in an inert atmosphere. This appears to occur as a result of switching the limiting step from random scission to decomposition of the hydroperoxide radical, which occurs with a lower activation energy [34]. The combustion of polystyrene takes place simultaneously with both pyrolysis and thermal oxidation, with the pyrolysis prevailing over thermal oxidation [62]. For the most part, styrene and toluene were present after its degradation at various temperatures under a vacuum [63] (Table 1.11).

1.5.6 Polylactic Acid (PLA)

Poly(lactic acid) is a biodegradable hydrolysable aliphatic semi-crystalline polyester produced through the direct condensation reaction of its monomer, lactic acid, as the oligomer, and followed by a ring-opening polymerisation of the cyclic lactide dimer [65]. It is derived from renewable resources, such as corn starch or sugarcanes, and is considered biodegradable and compostable. PLA is a thermoplastic, high-strength, high-modulus polymer that can be made from annually renewable sources

Table 1.11 Products of the thermal degradation of polystyrene

Thermal degradation in a vacuum[a]			
Temperature [°C]	300	350	420
Products	Styrene Toluene Benzene Naphthalene α-methylstyrene 1-methylindene 3-phenylpropene	Styrene Toluene α-methylstyrene 1-methylindene 3-phenylpropene Naphthalene Dimethylindene Trans-2-methylstyrene 3-methylindane 4-phenyl-1-butene	Styrene Toluene Ethylbenzene α-methylstyrene
Thermal degradation in air[b]			
Temperature [°C]	200	350	500
Products	Styrene Ethylbenzene Cumene	Styrene Benzaldehyde 1-phenylethanol Acetophenone Styrene oxide Phenol α-styrene Allylbenzene Ethylbenzene Benzene Toluene Cumene n-Propylbenzene Benzyl alcohol Cinnamaldehyde	Styrene Benzaldehyde Toluene 1-phenylethanol α-styrene Styrene oxide Benzoic acid Allylbenzene Acetophenone Phenol Benzene Ethylbenzene Cumene n-Propylbenzene Cinnamaldehyde Benzyl alcohol

[a] McNeill et al. [63]
[b] Pfäffli et al. [64]

to yield articles for use in either the industrial packaging field or the biocompatible/bioabsorbable medical device market [41]. Bacteria are used to extract the lactic acid from the base material, and then, it is polymerised to make PLA [66]. The chemical/structural formula of this polymer is illustrated in Fig. 1.8.

It was found that PLA maintains relatively good thermal stability up to 280 °C [67]. In the course of the thermogravimetric analysis of PLA, a single-stage degradation process was indicated [68, 69]. It occurs in the temperature range of 240–390 °C. It was also observed that the presence of a coloured pigment increased the stability of polylactic acid [70]. Sin and Tueen [65] suggest that the thermal decomposition of PLA commonly takes place between 230 and 260 °C. According to Wojtyła et al. [71], thermal decomposition occurs in the range of 300–400 °C. A single-stage degradation process is indicated with one DTG peak at 374 °C. They also observed the loss of the mass below 300 °C. Teoh et al. [72] indicated that the thermal decomposition of PLA

1.5 The Thermal Degradation of Selected Polymers

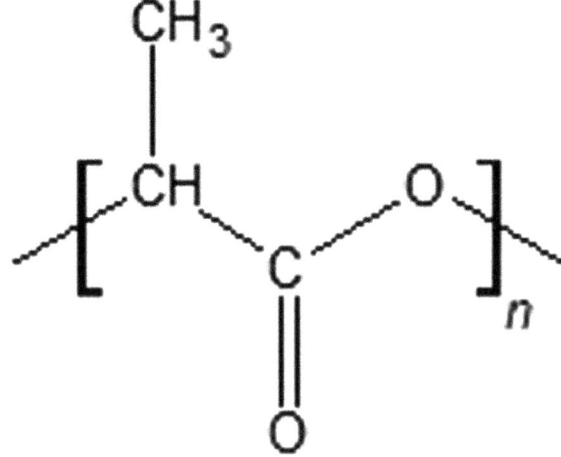

Fig. 1.8 Chemical formula of polylactic acid

in inert atmosphere began at up to 326 °C and reported the end of decomposition temperature at 380 °C. PLA thermally decomposed without leaving any residue. According to Karakoc et al. [73], the PLA filament starts to decompose, in air, at 321 °C. Although the decomposition occurs in two stages, the first stage already results in a 95% weight loss.

During the thermal degradation of PLA, hydroxyl end-initiated ester interchange processes and chain homolysis take place. Their products include cyclic oligomers, lactide, acetaldehyde, carbon monoxide, and carbon dioxide. At high temperatures, the amount of carbon monoxide released decreases [69]. Lv et al. [74] explain that the main products of pyrolysis and thermal oxidation are rather similar. They include carbonyl compounds, carbon dioxide, carbon monoxide, methane, and water. The emission of CO_2 for thermo-oxidative degradation is higher and shows two steps compared with that of pyrolysis. This could be due to more decomposition products that are transformed to CO_2 in the presence of oxygen, and the generated bio-oil and chars could be further oxidised to CO_2 at a higher temperature.

1.5.7 *Acrylonitrile Butadiene Styrene Copolymer (ABS)*

A copolymerisation of acrylonitrile, butadiene, and styrene (Fig. 1.9), using the influence of acrylonitrile component, allows an increase in the chemical resistance of polystyrene, and at the same time, it substantially increases its toughness through the influence of the butadiene component. The strength and solidity of the material were sufficiently well preserved [1]. The individual components can be mixed in the following range of ratios [41]:

- Acrylonitrile—15%–35%
- Butadiene—5%–30%

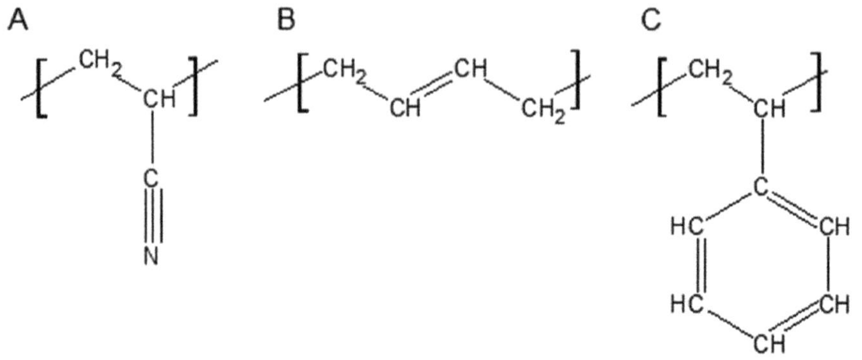

Fig. 1.9 The chemical formulae of ABS monomer units: A—acrylonitrile, B—butadiene, C—styrene

- Styrene—40%–60%.

ABS polymers are heterogeneous materials. In the continuous phase of the styrene-acrylonitrile copolymer, small particles of polybutadiene rubber are dispersed. These are characterised by their low resistance when exposed to inclement weather conditions and ageing when exposed to light [16]. ABS is resistant to aqueous acids, alkalis, concentrated hydrochloric and phosphoric acids, alcohols, and oils, but it swells when exposed to glacial acetic acid, carbon tetrachloride, and aromatic hydrocarbons and is attacked by concentrated sulfuric and nitric acids. It is soluble in esters, ketones, and ethylene dichloride [41].

According to Suzuki and Wilkie [75], the degradation of ABS in an inert gas occurs in a single step and it begins at 340 °C and leaves a 4% residue at 600 °C. Yang et al. [76] conducted a thermogravimetric analysis at a lower heating rate (10 °C min^{-1}). They state that the thermal degradation of ABS in nitrogen is a two-step process with the major weight loss happening in the first step. Compared to the degradation in air, the curve has longer and larger tails because without the inclusion of oxygen, the thermal degradation takes place more slowly. A two-stage degradation mechanism is also described by Klarić et al. [77], and it begins at approximately 360 °C when exposed to a heating rate of 2.5 °C min^{-1}.

In air, ABS degrades in two steps. The first step initiates at 180 °C and ends at 480 °C and the second step from 480 °C up to 620 °C. The first degradation step is not a simple chemical reaction but several reactions that occur simultaneously [76].

At the beginning of the ABS pyrolysis, a butadiene monomer is released. Aromatics are first noted at 350 °C, a temperature at which the evolution of butadiene is still evident. As the temperature is increased, styrene becomes more prevalent and at 420 °C the intensities of the C–H bands in butadiene and styrene are about equal. At higher temperatures, the aromatics decrease in intensity while butadiene remains very strong [75]. Thermal oxidation of ABS basically comes from its polybutadiene component as a result of hydroperoxide or carbonyl complexes. Oxidation of the polybutadiene segment phase in ABS leads to an exothermic and self-accelerating effect at moderate temperature, which can be a potential hazard in an ABS plant [78].

1.5.8 Polyethylene Terephthalate (PET) and Polyethylene Terephthalate Glycol (PETG)

Polyethylene terephthalate is a linear type of polyester, meaning it has a non-branched polymer chain [2]. In the past, it was primarily used for the production of fibres, but today it is more commonly used for plastic bottles for beverages. Its main properties include [79]:

- Transparency to visible light and microwaves
- Very good resistance to ageing, wear, and heat
- Lightweight, impact, and shatter resistant
- Good gas and moisture barrier properties.

PETG is a glycol-modified variant of polyethylene terephthalate. The addition of glycol results in a lower melting temperature and viscosity compared to PET and an amorphous structure in comparison with PET. Due to these properties, PETG is increasingly being used as a filament for 3D printers (Friedrich et al. [132]).

The chemical/structural formulae of both polymers are illustrated in Fig. 1.10.

The temperature at which decomposition of pure PET starts in nitrogen is 412 °C [81]. Karakoç et al. [73] describe the decomposition of a PET filament in air. They suggest it begins at 381 °C and divide its course into two stages. During the first stage, the weight loss was 84.24%, and during the second stage, it was 15.59%. While PET practically completely decomposes in an atmosphere that contains oxygen, in an inert atmosphere, the reactions almost completely stop when there is still a residue of 20% and then continue at a very slow rate. The value of the degradation temperature is lower for thermo-oxidative reactions in comparison with pyrolytic degradation, which implies that the thermal degradation of PET in an oxygenated environment occurs at a faster rate than the degradation in a pyrolytic inert condition [82]. The PETG samples appear to be a material with high thermal stability. Under a nitrogen heat flow, the degradation process begins above 380 °C and occurs in a single step. Although the initial stage of the decomposition of PET and PETG is very similar, by adding 1,4-cyclohexanedimathenol terephthalate the rate of thermal degradation increases and the amount of residue after pyrolysis is reduced [83]. The PETG mass loss is greater than 90% from room temperature to 650 °C, and its thermal decomposition is mainly complete in a single step at 425 °C.

A thermogravimetric analysis of PETG in air reveals two-stage decomposition, just as in the case of PET [84].

As Ohtani et al. [85] suggest, a large amount of benzoic acid is released during the pyrolysis of PET. Other products include monoalkenyl esters of terephthalic acid, alkenyl esters of benzoic acid, mono- and dialkenyl esters related to two terephthalic

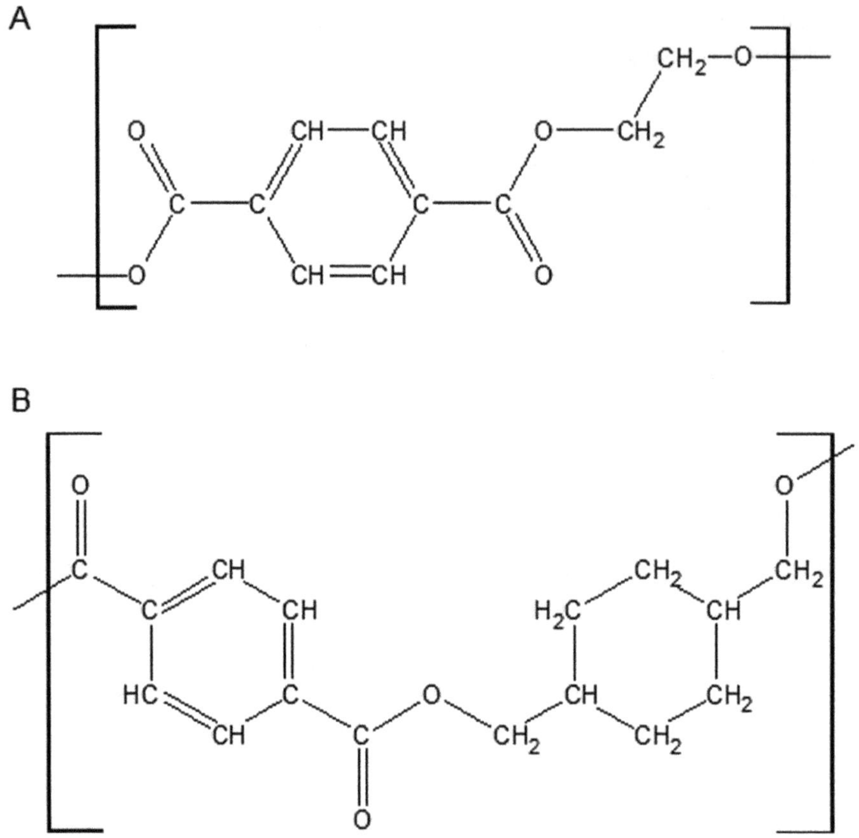

Fig. 1.10 Chemical formulae of the structural units of PET and PETG: **a** ethylene terephthalate unit, **b** 1,4-cyclohexanedimethanol terephthalate unit (according to Chen et al. [80])

acid units, benzene, biphenyl, dialkenyl esters of terephthalic acid and dibenzoates, and 19% of minor by-products with shorter retention times than that of benzene. During the thermal degradation of PET in air at 500 °C, FTIR identified carbon oxides and water, along with organic substances including methane, ethane, ethyne, and formaldehyde. PET combustion at 800 °C is a more efficient burning process. The main combustion products are carbon oxides and water. The characteristic groups we find in a GC–MS analysis are aromatic compounds, derivates of benzoic acid, a group of phthalates, biphenyles, and others [86].

Yu and Huang state that the aliphatic backbone in PETG plays a dominant role in determining the behaviour of pyrolysis. 4-methylene-cyclohexanemethanol, 1,4-bis-(methylene)-cyclohexane, and benzoic acid are characteristic thermal decomposition products [87].

1.5.9 Natural Polymers

Natural polymers are derived by extraction from their natural bulk form, for example, cellulose or lignin is extracted from wood. This also includes polymers produced by biological processes such as bacterial synthesis or fermentation. They may be divided into six main groups [88]:

1. Proteins
2. Polysaccharides
3. Polynucleotides
4. Polyisoprenes
5. Polyesters
6. Lignin.

Natural polymers that contain cellulose or possibly lignin are the most vulnerable to fire. This group includes wood or paper. As to composition, they are mostly made up of carbon and oxide, which accounts for up to 90% of their weight. They also contain rather remarkable amounts of hydrogen. Due to its low atomic weight, calculations show that it only represents 6 w/w %. However, it is the most prevalent element, representing up to 45% of the total material. In addition to the above-mentioned elements, lignocellulose materials may also contain trace amounts of other elements, especially nitrogen and sulphur. A comparison of the elemental composition of selected lignocellulose materials is provided in Table 1.12.

Cellulose is a natural polysaccharide, and its molecule consists of D-glucose units mutually bound by β-$(1\rightarrow 4)$-glycosidic bonds (Fig. 1.11). It is the most common polymer material found in nature. In its most common form, it is a tough, fibrous, water-insoluble material that is mostly found in the cell walls of plants, mainly in the stalks, stems, or trunks [19]. In the animal realm, cellulose is synthesised by tunicates [103, 104].

The first reaction that occurs in the thermal decomposition of cellulose is the modification of cellulose into "active cellulose". This was first described by Bradbury et al. [105]. This idea is based on the initial induction period that is observed during thermogravimetric analysis of cellulose. Matsuoka et al. [106] proposed a mechanism for active cellulose: the thermal decomposition of reducing end groups, which are originally present or are formed during the pyrolysis of crystalline cellulose, then activates the following pyrolysis reactions.

The pyrolysis of cellulose occurs in two stages [107: 435–444]:

1. By dehydration and subsequent charring (intra-ring scission of the glucose unit in the cellulose chain)
2. Transglycosylation and levoglucosan formation.

Shafzadeh [108] described this classification as early as 1968. He stated that the thermal degradation of cellulose may occur along two main pathways. One involves fragmentation and the formation of combustible volatiles which could feed the flames, and the second mainly involves dehydration and the formation of carbonaceous char that could lead to localised, and relatively slower, glowing ignition. While

Table 1.12 Elemental composition of selected lignocellulose materials

Material	C [%hm.]	H [%hm.]	O [%hm.]	N [%hm.]	S [%hm.]	Source
Larch	46.92	6.73	46.23	0.12	–	[89]
Beech	46.9	6.2	45.9	0.3	–	[90]
Spruce	48.3	6.3	44.6	0.4	–	[90]
Iroko	43.9	5.3	46.9	0.4	–	[90]
Albizia	46.4	5.8	45.5	0.6	–	[90]
Corn cob	43.6	5.8	48.6	0.7	–	[90]
Beech	45.52	6.34	47.98	0.16	–	[91]
Pine	50.35	6.33	38.74	0.17	0.01	[92]
Birch	49.53	6.26	39.85	0.19	0.07	[92]
Alkali lignin	62.4 ± 0.14	6.14 ± 0.00	29.43 ± 0.03	0.26 ± 0.03	1.77 ± 0.06	[93]
Hemicellulose	38.03	5.31	43.1	0.03	0	[94]
Cellulose	39.63	5.31	49.88	0	0	[94]
Lignin	61.45	5.54	23.98	0.92	1.5	[94]
Kraft Lignin	69.3	6.2	25.8	0.8	1.7	[95]
Kraft lignin	65.2 ± 0.2	6.1 ± 0.2	27.4 ± 0.9	0.1 ± 0.05	0.8 ± 0.2	[96]
Cellulose	43	6.3	50.7	–	–	[97]
Hemicellulose	43.88	6.48	43.11	0.72	–	[98]
Cellulose	44.55	6.62	48.31	0.52	–	[98]
Lignin	64.0	5.83	24.4	1.77	–	[98]
Cellulose	44.06	5.98	49.49	0.41	0.06	[99]
Xylan	40.25	5.76	53.62	0.32	0.05	[99]
Lignin	58.70	5.45	35.43	0.37	0.05	[99]
Poplar	43.1	5.4	51.5	–	–	[100]
Cellulose	40.9	6.4	52.7	–	–	[100]
Lignin	48.3	5.1	46.6	–	6.4	[100]
Beech	50.8	5.9	42.9	0.3	0.02	[100]
Eastern redcedar	51.07	5.97	40.95	0.37	0.0	[101]
Switchgrass	46.62	5.74	42.27	0.18	0.3	[101]
Wheat straw	43.2	5.0	39.4	0.61	0.11	[101]
Palm fibre	62.0	9.1	27.4	1.3	0.2	[102]

dehydration and charring reactions prevail at lower temperatures, the formation of levoglucosan takes place at somewhat higher temperatures. The general pathways for the pyrolysis of cellulose, leading to the production of char as well as gaseous and volatile products, are shown in Fig. 1.12. While inorganic additives move the

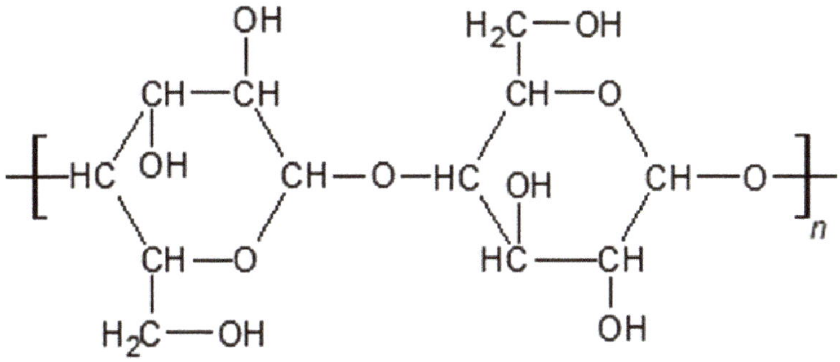

1.5 The Thermal Degradation of Selected Polymers

Fig. 1.11 The chemical formula of cellulose

burning towards smouldering, rapid air flow with a high intensity heat flux creates better conditions for flaming [109].

According to Lin et al. [97], the pyrolysis of cellulose may be briefly described as follows:

1. Cellulose depolymerises while producing levoglucosan;

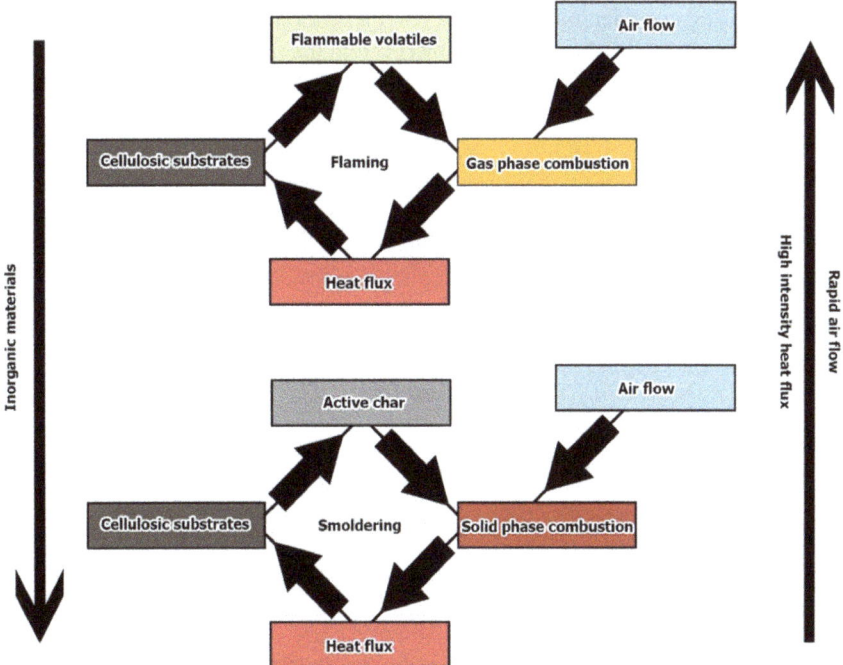

Fig. 1.12 A schematic depiction of flaming and smouldering combustion (processed based on Shafizadeh [109])

2. Levoglucosan undergoes dehydration and isomerisation reactions to form other anhydro-monosaccharides including levoglucosenone, 1,4:3,6-dianhydro-β-D-glucopyranose, and 1,6-anhydro-β-D-glucofuranose;
3. The anhydrosugars react to form furans, such as furfural and hydroxymethylfurfural, by dehydration reactions or hydroxyacetone, glycolaldehyde, and glyceraldehyde by fragmentation and retroaldol condensation reactions;
4. Carbon monoxide and carbon dioxide are formed from decarbonylation and decarboxylation reactions;
5. Char is formed from polymerisation of the pyrolysis products.

Paulsen et al. [110] suggest it is likely that furans do not form via anhydrosugars but rather directly from the cellulose chain. They also show that once the furan ring is formed, it does not breakdown.

Thermogravimetric measurements of the decomposition of cellulose in an inert atmosphere revealed a major mass loss stage that may be attributed to pyrolytic decomposition. After adding oxygen, another stage was observed, triggered by the oxidation of the carbon residue. The temperature range in which the second stage occurred, as well as the weight of the residue after the measurements, was affected by the concentration of oxygen (Table 1.13) [111, 112].

Oxygen at low temperatures only interacts with surface cellulose. Only about 3% of the cellulose is gasified as a result of the oxidation in the early stages of decomposition preceding considerable thermal depolymerisation. Oxygen intensively reacts with the products of the thermal depolymerisation of cellulose, although it does not penetrate into the matrix of polymer cellulose. In other words, the oxidation is a fast process subsequent to thermal depolymerisation [113].

The term Lignin was introduced in 1819 by Augustin de Candolle and originates from the Latin word lignum, meaning wood [114]. Lignin does not refer to a single chemical substance. It may occur in various forms, and often even derivates of the original natural lignin are called by the same name. This is why scientific literature

Table 1.13 A comparison of the TGA of cellulose in atmospheres with various oxygen contents [111, 112]

Atmosphere [% O_2]	Temperature range [°C]	Mass loss range [%]
First mass loss stage		
0	270–420	98.87–8.86
7	270–400	99.10–9.53
20	270–395	99.07–10.21
60	270–390	98.62–14.09
Second mass loss stage		
7	400–610	9.53–2.04
20	395–590	10.21–1.43
60	390–565	14.09–1.39

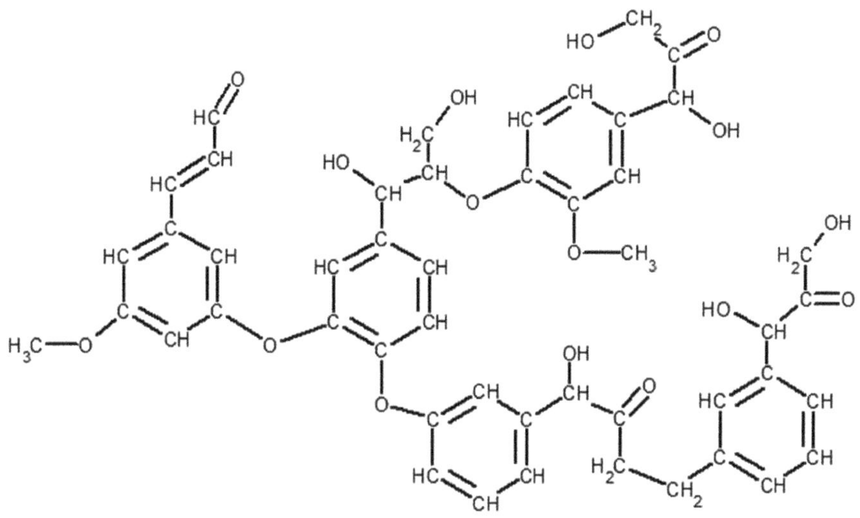

1.5 The Thermal Degradation of Selected Polymers

Fig. 1.13 An example of the chemical formula of lignin (based on Sharma et al. [116])

equally utilises the term lignins. One of the possible chemical/structural formulae of lignin is shown in Fig. 1.13.

The term lignin tends to be defined differently depending on the needs of the field, where the definition is employed. It is also classified based on the method of isolation used. The most common found is Klason lignin, but there is also hydrolytic lignin, alkali lignin, and others. According to widely accepted convention, lignin may be defined as an amorphous, polyphenolic material that arises from an enzyme-mediated dehydrogenative polymerisation of three phenylpropanoid monomers, coniferyl, sinapyl, and p-coumaryl alcohols [115].

Lignin thermally degrades at a rather wide temperature range because the oxygen-containing functional groups in its structure have different thermal stabilities, and therefore, they break down at various temperatures. Considering its complex composition and structure, the thermal degradation of lignin is substantially affected by its nature, moisture content, reaction temperature, the surrounding atmosphere, and the processes of heat and substance transfer [117]. Its thermal stability is also dependent on the method of isolation [118].

Jiang, Nowakowski, and Bridgwater studied the impact of temperature on the composition of the by-products of the pyrolysis of lignin. The following conclusions were drawn based on the measured results [119]:

- The product distribution from lignin pyrolysis depends upon the pyrolysis temperature. The maximum yield of phenolic compounds was obtained at 600 °C;
- At higher temperatures, demethylation, demethoxylation, decarboxylation, and alkylation occur, leading to the change of by-product distribution towards alkyl-phenol and polyhydroxybenzene;
- The yields of most of the degradation compounds are less than 1%.

In the same year, they also published an article in which they state that volatiles and char production depends on the lignin separation method. Organosolv lignin, hydrolytic lignin, and alkali lignin have similar volatiles production, which is greater than that from Klason lignin [119]. In the initial stages of the degradation of lignin, endothermic phenomena occur due to the release of moisture and adsorbed gases, followed by softening. During pyrolysis, three successive exothermic reactions occur [120]:

- At 280 °C, where scission of aliphatic groups begins and some lignin on the surface begins to carbonise
- At 380 °C, where scission of aromatic parts occurs and all the lignin on the surface is carbonised
- At 460 °C, where the carbon in the char is condensed into graphitelike rings.

During the thermogravimetric analysis, lignin decomposes, in an inert atmosphere, in a single step (pyrolysis) and in an atmosphere that contains oxygen in two steps (pyrolysis + char oxidation). The first mass loss stage is not significantly influenced by the atmosphere in comparison with the second mass loss stage (Table 1.14). The evolution profiles of CO and CO_2 match the DTG curves very well regardless of the atmosphere. Their formation is closely related to the carbon complexes on the char surface, which is mainly controlled by the chemical-sorption progress [111, 112].

Wood

Wood is one of the main raw materials used by humans. From a chemical perspective, it is mostly composed of three components: hemicellulose, cellulose, and lignin. Hemicelluloses and cellulose create a carbohydrate section. This is usually referred to as holocellulose. In addition to the main components, there are also other accompanying substances present in wood. Some of them are soluble in various solvents, which is why they are referred to as extractives. Inorganic materials are called ashes.

Table 1.14 A comparison of the TGA of lignin in atmospheres with various oxygen contents [111, 112]

Atmosphere [% O_2]	Temperature range [°C]	Mass loss range [%]
First mass loss stage		
0	220–550	97.28–41.06
7	220–480	97.69–53.96
20	220–455	98.86–59.91
60	220–437	98.27–62.05
Second mass loss stage		
7	480–670	53.96–3.61
20	455–630	59.91–3.98
60	437–580	62.05–3.38

1.5 The Thermal Degradation of Selected Polymers

Table 1.15 Composition of various wood species

Material	Extractives	Hemicellulose	Cellulose	Lignin	Source
Larch	2.83	23.34	46.45	22.40	[89]
Hornbeam	7.7	23.3	48.9	20.1	[122]
Walnut	4.1	25.9	22.1	47.8	[122]
Scots pine	2.7	20.8	49.8	26.7	[122]
Pine	8.7	17.7	47.8	25.5	[92]
Birch	7.59	28.97	53.95	9.43	[92]
Poplar wood	3.09	28.22	52.99	15.70	[123]
Spruce	1.1 ± 0.4	22.3 ± 0.9	47.1 ± 0.4	29.2 ± 0.6	[124]
Oak	1.6 ± 0.3	21.9 ± 0.7	43.2 ± 0.3	35.4 ± 0.4	[124]
Pine	5.7 ± 0.1	24.0 ± 0.7	45.6 ± 0.1	26.8 ± 0.3	[124]

The representation of organic elements in wood is indicated in Table 1.15. Cellulose is the most prevalent component.

Overall, dry wood has an elemental composition of about 50% carbon, 6% hydrogen, 44% oxygen, and trace amounts of inorganics. In general, coniferous species have a higher cellulose content, higher lignin, and lower pentosan content when compared to deciduous species [121].

Poletto et al. [125] studied the impact of the individual components of wood on its thermal degradation. They suggest that a higher content of extractive substances accelerates the process of degradation. The higher reactivity of hemicelluloses and lignin may accelerate degradation reactions and result in the degradation of cellulose under relatively low temperatures. On the other hand, organised cellulose regions suppress degradation. The hemicellulose components in softwoods show a lower reactivity. Softwoods also include a wider zone of cellulose decomposition. Char yields are 14–23% for hardwoods and 20–26.5% for softwoods [126]. At low heating rates, volatile pyrolysis products are released through the natural porosity. On the contrary, at high heating rates, the original cellular structure is lost, as a consequence of melting phenomena. Fast volatile release produces substantial internal overpressure and coalescence of the smaller pores, leading to large internal cavities and a more open structure of both wood and lignin. Hence, for pyrolysis carried out at atmospheric pressure, chars produced at low heating rates mainly consist of a micropore structure, whereas those obtained with high heating rates mainly present macropores [127]. Müller-Hagedorn et al. [122] explain that the thermal degradation of wood is also affected by the content of mineral salts, which have a strong effect on the pyrolysis temperature and therefore on the pyrolysis kinetics and mechanisms. The effect of higher temperatures on wood is shown in Table 1.16.

In the case of the thermal degradation of wood (beech), it was determined that most of the gaseous products were released in the temperature range of 500–700 K. A substantial amount of formaldehyde was produced, which was almost triple the

Table 1.16 Temperature range of wood pyrolysis and combustion [128]

Temperature range	Decomposition processes
>100 °C	The evaporation of chemically unbound water
160–200 °C	The three-polymeric components of wood begin to slowly decompose. Gases formed at this stage are non-combustible (mainly H_2O)
200–225 °C	Wood pyrolysis is still very slow, and most of the gases produced are non-combustible
225–275 °C	The main pyrolysis begins and flaming combustion will occur with the aid of a pilot flame
280–500 °C	Gases produced are now volatile (CO, methane etc.) and smoke particles are visible. Char forms rapidly as the physical structure breaks down
>500 °C	Volatile production is complete. Char continues to smoulder and oxidise to the forms CO, CO_2, and H_2O

amount of methanol and ten times the amount of methane at their respective peaks [91].

Thermal degradation of wood in an atmosphere containing oxygen differs from degradation in inert gases in a similar way to the polymer materials it is composed of. In addition to a shift in the individual phases of degradation towards lower temperatures, oxygen also causes oxidation of the carbonised residue. These differences are clearly demonstrated in the data of [127] shown in Table 1.17.

A proximate analysis provides crucial information with regard to the combustion of lignocellulose materials. It determines how much of the sample, during pyrolysis, produces volatile compounds, how much is composed of carbonised structure and how much of inflammable residue (ash). Volatile by-products are especially important for flaming combustion, and fixed carbon for the thermal oxidation on the interface of the gas and solid phases. At the same time, the fixed carbon content of the samples is an effective indicator of the ability to produce char from a given wood [126]. Cellulose contains more volatile matter (more than 90%), whereas lignin has a significantly higher content of fixed carbon (25–45%). A proximate analysis of various lignocellulosic materials is indicated in Table 1.18.

Table 1.17 Phases of the thermal degradation of various wood species in air and an inert atmosphere (argon) [127]

Sample	Temperature ranges [°C]					
	Drying		Devolatilisation		Char combustion	
	Air	Ar	Air	Ar	Air	
Oak	<104	<118	209–349	219–377	362–517	
Pine	<96	<120	215–344	224–378	355–512	
Spruce	<92	<120	210–340	221–377	347–503	
Aspen	<91	<120	209–345	222–369	351–509	

1.5 The Thermal Degradation of Selected Polymers

Table 1.18 Proximate analysis of lignocellulosic materials

Material	Volatile matters	Fixed carbon	Ash	Source
Redwood	82.3	17.5	0.2	[126]
Spruce	84.4	14.9	0.7	[126]
Lignin	49.4–54.6	45.2–47.9	0.2–2.7	[126]
Douglas fir	84.2	15.4	0.3	[126]
Pine	85	14.7	0.3	[126]
Pine	86.6	13.1	0.3	[126]
Alder	86	13.7	0.3	[126]
Beech	86.5	13.1	0.4	[126]
Birch	87.4	12.4	0.2	[126]
Oak	84.4	15.5	0.1	[126]
Beech	85.9	13.4	0.7	[90]
Spruce	83.5	16.1	0.4	[90]
Iroko	70.4	26.1	3.5	[90]
Albizia	72.2	25.5	1.8	[90]
Corn cob	80.6	18.2	1.2	[90]
Pine	82.47	17.28	0.25	[92]
Birch	86.82	12.89	0.29	[92]
Hemicellulose	82.16	14.67	3.17	[94]
Cellulose	94.56	5.22	0.22	[94]
Alkali lignin	66.43 ± 0.21	27.36 ± 0.28	6.21 ± 0.07	[93]
Lignin	66.02	31.24	2.74	[94]
Cellulose	94.8	5.1	0.1	[97]
Hemicellulose	82.69	11.48	5.82	[98]
Cellulose	97.89	2.11	–	[98]
Cellulose	95.12	4.66	0.22	[99]
Xylan	81.46	13.17	5.37	[99]
Lignin	72.05	24.79	3.16	[99]
Lignin	69.61	26.40	3.99	[98]
Palm fibre	89.8	10.1	0.1	[102]
Beech	85.3	14.3	0.4	[129]
Spruce	89.35	10.25	0.4	[124]
Oak	90.24	9.56	0.2	[124]
Pine	88.54	10.35	1.11	[124]

References

1. Ducháček V (2006) Polymery - výroba, vlastnosti, zpracování, použití, Praha
2. Masařík I (2003) Plasty a jejich požární nebezpečí, SPBI Spektrum
3. Plastics Europe. Plastics—the facts 2011
4. Plastics Europe. Plastics—the facts 2012
5. Plastics Europe. Plastics—the facts 2013
6. Plastics Europe. Plastics—the facts 2014
7. Plastics Europe. Plastics—the facts 2015
8. Plastics Europe. Plastics—the facts 2016
9. Plastics Europe. Plastics—the facts 2017
10. Plastics Europe. Plastics—the facts 2018
11. Plastics Europe. Plastics—the facts 2019
12. Plastics Europe. Plastics—the facts 2020
13. Subramanian MN (2013) Plastics additives and testing. Wiley, Hoboken
14. Pritchard G (2005) Plastics additives: a Rapra market report. iSmithers Rapra Publishing
15. Gupta AL (2010) Polymer chemistry. Pragati Publications, Meerut, India
16. Mleziva J, Šňupárek J (2000) Polymery–výroba, struktura, vlastnosti a použití, 2. přepracované vydání. Sobotáles, Praha
17. Ram A (1997) Fundamentals of polymer engineering. Technion-Israel Institute of Technology, Haifa
18. Muccio EA (1994) Plastics processing technology. ASM International
19. Modjarrad K, Ebnesajjad S (eds) (2013) Handbook of polymer applications in medicine and medical devices. Elsevier, Amsterdam
20. Eaves D (2004) Handbook of polymer foams. Rapra Technology Ltd., Shrewsbury
21. Ray SS, Kuruma M (2019) Halogen-free flame-retardant polymers next-generation fillers for polymer nanocomposite applications
22. Shrivastava A (2018) Introduction to plastics engineering. William Andrew, Norwich
23. Xie D, Zhao Y, Li Y, LaChance AM, Lai J, Sun L, Chen J (2019) Rheological, thermal, and degradation properties of PLA/PPG blends. Materials 12(21):3519
24. Callister Jr WD, Rethwisch DG (2020) Fundamentals of materials science and engineering: an integrated approach. Wiley, Hoboken
25. Pielichowski K, Njuguna J (2005) Thermal degradation of polymeric materials. iSmithers Rapra Publishing
26. Johnson PR (1974) A general correlation of the flammability of natural and synthetic polymers. J Appl Polym Sci 18(2):491–504
27. Tian C, Shi Z, Zhang H, Xu J, Shi J, Guo H (1999) Thermal degradation of cotton cellulose. J Therm Anal Calorim 55(1):93–98
28. Chatterjee A, Deopura BL (2006) Thermal stability of polypropylene/carbon nanofiber composite. J Appl Polym Sci 100(5):3574–3578
29. Hornsby PR, Watson CL (1990) A study of the mechanism of flame retardance and smoke suppression in polymers filled with magnesium hydroxide. Polym Degrad Stab 30(1):73–87
30. Wang Z, Qu B, Fan W, Huang P (2001) Combustion characteristics of halogen-free flame-retarded polyethylene containing magnesium hydroxide and some synergists. J Appl Polym Sci 81(1):206–214
31. Wei LL, Wang DY, Chen HB, Chen L, Wang XL, Wang YZ (2011) Effect of a phosphorus-containing flame retardant on the thermal properties and ease of ignition of poly (lactic acid). Polym Degrad Stab 96(9):1557–1561
32. Heskestad G (1998) Dynamics of the fire plume. Philos Trans Roy Soc Lond Ser A Math Phys Eng Sci 356(1748):2815–2833
33. Hilado CJ (1998) Flammability handbook for plastics. CRC Press, Boca Raton
34. Peterson JD, Vyazovkin S, Wight CA (2001) Kinetics of the thermal and thermo-oxidative degradation of polystyrene, polyethylene and poly (propylene). Macromol Chem Phys 202(6):775–784

References

35. Tsuchiya Y, Sumi K (1969) Thermal decomposition products of polypropylene. J Polym Sci Part A-1: Polym Chem 7(7):1599–1607
36. Audisio G, Silvani A, Beltrame PL, Carniti P (1984) Catalytic thermal degradation of polymers: degradation of polypropylene. J Anal Appl Pyrol 7(1–2):83–90
37. Ballice L, Reimert R (2002) Classification of volatile products from the temperature-programmed pyrolysis of polypropylene (PP), atactic-polypropylene (APP) and thermogravimetrically derived kinetics of pyrolysis. Chem Eng Process 41(4):289–296
38. Purohit V, Orzel RA (1988) Polypropylene: a literature review of the thermal decomposition products and toxicity. J Am Coll Toxicol 7(2):221–242
39. ISO 17855-1:2014. Plastics—polyethylene (PE) moulding and extrusion materials—part 1: designation system and basis for specifications
40. Vasile C, Pascu M (2005) Practical guide to polyethylene. iSmithers Rapra Publishing
41. McKeen LW (2016) Permeability properties of plastics and elastomers. William Andrew, Norwich
42. Jordan JL, Casem DT, Bradley JM, Dwivedi AK, Brown EN, Jordan CW (2016) Mechanical properties of low density polyethylene. J Dyn Behav Mater 2(4):411–420
43. Beyler CL, Hirschler MM (2002) Thermal decomposition of polymers. SFPE handbook of fire protection engineering
44. Ueno T, Nakashima E, Takeda K (2010) Quantitative analysis of random scission and chain-end scission in the thermal degradation of polyethylene. Polym Degrad Stab 95(9):1862–1869
45. Zong R, Wang Z, Liu N, Hu Y, Liao G (2005) Thermal degradation kinetics of polyethylene and silane-crosslinked polyethylene. J Appl Polym Sci 98(3):1172–1179
46. Faravelli T, Bozzano G, Colombo M, Ranzi E, Dente M (2003) Kinetic modeling of the thermal degradation of polyethylene and polystyrene mixtures. J Anal Appl Pyrol 70(2):761–777
47. Park JW, Oh SC, Lee HP, Kim HT, Yoo KO (2000) A kinetic analysis of thermal degradation of polymers using a dynamic method. Polym Degrad Stab 67(3):535–540
48. Dubdub I, Al-Yaari M (2020) Pyrolysis of mixed plastic waste: I. Kinetic study. Materials 13(21):4912
49. Bockhorn H, Hornung A, Hornung U, Schawaller D (1999) Kinetic study on the thermal degradation of polypropylene and polyethylene. J Anal Appl Pyrol 48(2):93–109
50. Hodgkin JH, Galbraith MN, Chong YK (1982) Combustion products from burning polyethylene. J Macromol Sci Chem 17(1):35–44
51. Font R, Aracil I, Fullana A, Conesa JA (2004) Semivolatile and volatile compounds in combustion of polyethylene. Chemosphere 57(7):615–627
52. Ballice L, Yüksel M, Sağlam M, Reimert R, Schulz H (1998) Classification of volatile products from the temperature-programmed pyrolysis of low-and high-density polyethylene. Energy Fuels 12(5):925–928
53. Intratec (2019) Polyvinyl chloride via emulsion polymerization process. Report PVC E13A, 63 p
54. Patrick S (2005) Practical guide to polyvinyl chloride. iSmithers Rapra Publishing
55. Xu ZP, Saha SK, Braterman PS, D'Souza N (2006) The effect of Zn, Al layered double hydroxide on thermal decomposition of poly (vinyl chloride). Polym Degrad Stab 91(12):3237–3244
56. McNeill IC, Memetea L, Cole WJ (1995) A study of the products of PVC thermal degradation. Polym Degrad Stab 49(1):181–191
57. Tsuchiya Y, Sumi K (1967) Thermal decomposition products of polyvinyl chloride. J Appl Chem 17(12):364–366
58. Szycher M (ed) (1999) Szycher's handbook of polyurethanes. CRC Press, Boca Raton
59. Rahman MM, Rabbani MM, Saha JK (2019) Polyurethane and its derivatives. In: Jafar Mazumder M, Sheardown H, Al-Ahmed A (eds) Functional polymers. Polymers and polymeric composites: a reference series. Springer, Berlin
60. Montaudo G, Puglisi C, Scamporrino E, Vitalini D (1984) Mechanism of thermal degradation of polyurethanes. Effect of ammonium polyphosphate. Macromolecules 17(8):1605–1614

61. Chattopadhyay DK, Webster DC (2009) Thermal stability and flame retardancy of polyurethanes. Prog Polym Sci 34(10):1068–1133
62. Rossi M, Camino G, Luda M (2001) Characterisation of smoke in expanded polystyrene combustion. Polym Degrad Stab 74(3):507–512
63. McNeill IC, Zulfiqar M, Kousar T (1990) A detailed investigation of the products of the thermal degradation of polystyrene. Polym Degrad Stab 28(2):131–151
64. Pfäffli P, Zitting A, Vainio H (1978) Thermal degradation products of homopolymer polystyrene in air. Scand J Work Environ Health 22–27
65. Sin LT, Tueen BS (2019) Polylactic acid: a practical guide for the processing, manufacturing, and applications of PLA. William Andrew, Norwich
66. Ashby MF, Johnson K (2013) Materials and design: the art and science of material selection in product design. Butterworth-Heinemann, Oxford
67. Mróz P, Białas S, Mucha M, Kaczmarek H (2013) Thermogravimetric and DSC testing of poly (lactic acid) nanocomposites. Thermochim Acta 573:186–192
68. Yang MH, Lin YH (2009) Measurement and simulation of thermal stability of poly (lactic acid) by thermogravimetric analysis. J Test Eval 37(4):364–370
69. Zou H, Yi C, Wang L, Liu H, Xu W (2009) Thermal degradation of poly (lactic acid) measured by thermogravimetry coupled to Fourier transform infrared spectroscopy. J Therm Anal Calorim 97(3):929–935
70. Matos BDM, Rocha V, da Silva EJ, Moro FH, Bottene AC, Ribeiro CA et al (2019) Evaluation of commercially available polylactic acid (PLA) filaments for 3D printing applications. J Therm Anal Calorim 137(2):555–562
71. Wojtyła S, Klama P, Baran T (2017) Is 3D printing safe? Analysis of the thermal treatment of thermoplastics: ABS, PLA, PET, and nylon. J Occup Environ Hyg 14(6):D80–D85
72. Teoh EL, Mariatti M, Chow WS (2016) Thermal and flame resistant properties of poly (lactic acid)/poly (methyl methacrylate) blends containing halogen-free flame retardant. Procedia Chem 19:795–802
73. Karakoç A, Rastogi VK, Isoaho T, Tardy B, Paltakari J, Rojas OJ (2020) Comparative screening of the structural and thermomechanical properties of FDM filaments comprising thermoplastics loaded with cellulose, carbon and glass fibers. Materials 13(2):422
74. Lv S, Zhang Y, Tan H (2019) Thermal and thermo-oxidative degradation kinetics and characteristics of poly (lactic acid) and its composites. Waste Manage 87:335–344
75. Suzuki M, Wilkie CA (1995) The thermal degradation of acrylonitrile-butadiene-styrene terpolymei as studied by TGA/FTIR. Polym Degrad Stab 47(2):217–221
76. Yang S, Castilleja JR, Barrera EV, Lozano K (2004) Thermal analysis of an acrylonitrile–butadiene–styrene/SWNT composite. Polym Degrad Stab 83(3):383–388
77. Klarić I, Roje U, Bravar M (1996) Thermooxidative degradation of poly (vinyl chloride)/acrylonitrile–butadiene–styrene blends. J Appl Polym Sci 61(7):1123–1129
78. Duh YS, Ho TC, Chen JR, Kao CS (2010) Study on exothermic oxidation of acrylonitrile-butadiene-styrene (ABS) resin powder with application to ABS processing safety. Polymers 2(3):174–187
79. Crawford CB, Quinn B (2017) Physiochemical properties and degradation. In: Microplastic pollutants, pp 57–100
80. Chen T, Zhang W, Zhang J (2015) Alkali resistance of poly (ethylene terephthalate)(PET) and poly (ethylene glycol-co-1, 4-cyclohexanedimethanol terephthalate)(PETG) copolyesters: the role of composition. Polym Degrad Stab 120:232–243
81. Xia XL, Liu WT, Tang XY, Shi XY, Wang LN, He SQ, Zhu CS (2014) Degradation behaviors, thermostability and mechanical properties of poly (ethylene terephthalate)/polylactic acid blends. J Central South Univ 21(5):1725–1732
82. Das P, Tiwari P (2019) Thermal degradation study of waste polyethylene terephthalate (PET) under inert and oxidative environments. Thermochim Acta 679:178340
83. Chen T, Jiang G, Li G, Wu Z, Zhang J (2015) Poly (ethylene glycol-co-1, 4-cyclohexanedimethanol terephthalate) random copolymers: effect of copolymer composition and microstructure on the thermal properties and crystallization behavior. RSC Adv 5(74):60570–60580

84. Paszkiewicz S, Irska I, Piesowicz E (2020) Environmentally friendly polymer blends based on post-consumer glycol-modified poly (ethylene terephthalate)(PET-G) foils and poly (ethylene 2, 5-furanoate)(PEF): preparation and characterization. Materials 13(12):2673
85. Ohtani H, Kimura T, Tsuge S (1986) Analysis of thermal degradation of terephthalate polyesters by high-resolution pyrolysis-gas chromatography. Anal Sci 2(2):179–182
86. Sovová K, Ferus M, Matulkova I, Španěl P, Dryahina K, Dvořák O, Civiš S (2008) A study of thermal decomposition and combustion products of disposable polyethylene terephthalate (PET) plastic using high resolution Fourier transform infrared spectroscopy, selected ion flow tube mass spectrometry and gas chromatography mass spectrometry. Mol Phys 106(9–10):1205–1214
87. Yu A, Huang XA (2006) Pyrolysis of an amorphous copolyester. J Appl Polym Sci 101(5):2793–2797
88. Olatunji O (ed) (2015) Natural polymers: industry techniques and applications. Springer, Berlin
89. Wang Z, Yang X, Sun B, Chai Y, Liu J, Cao J (2016) Effect of vacuum heat treatment on the chemical composition of larch wood. BioResources 11(3):5743–5750
90. Azeez AM, Meier D, Odermatt J, Willner T (2010) Fast pyrolysis of African and European lignocellulosic biomasses using Py-GC/MS and fluidized bed reactor. Energy Fuels 24(3):2078–2085
91. Ding Y, Ezekoye OA, Lu S, Wang C (2016) Thermal degradation of beech wood with thermogravimetry/Fourier transform infrared analysis. Energy Convers Manage 120:370–377
92. Shen DK, Gu S, Jin B, Fang MX (2011) Thermal degradation mechanisms of wood under inert and oxidative environments using DAEM methods. Biores Technol 102(2):2047–2052
93. Wang WL, Ren XY, Li LF, Chang JM, Cai LP, Geng J (2015) Catalytic effect of metal chlorides on analytical pyrolysis of alkali lignin. Fuel Process Technol 134:345–351
94. Qu T, Guo W, Shen L, Xiao J, Zhao K (2011) Experimental study of biomass pyrolysis based on three major components: hemicellulose, cellulose, and lignin. Ind Eng Chem Res 50(18):10424–10433
95. Caballero JA, Font R, Marcilla A (1996) Study of the primary pyrolysis of kraft lignin at high heating rates: yields and kinetics. J Anal Appl Pyrol 36(2):159–178
96. Yan Q, Li J, Zhang J, Cai Z (2018) Thermal decomposition of kraft lignin under gas atmospheres of argon, hydrogen, and carbon dioxide. Polymers 10(7):729
97. Lin YC, Cho J, Tompsett GA, Westmoreland PR, Huber GW (2009) Kinetics and mechanism of cellulose pyrolysis. J Phys Chem C 113(46):20097–20107
98. Zhao C, Jiang E, Chen A (2017) Volatile production from pyrolysis of cellulose, hemicellulose and lignin. J Energy Inst 90(6):902–913
99. Fan Y, Cai Y, Li X, Jiao L, Xia J, Deng X (2017) Effects of the cellulose, xylan and lignin constituents on biomass pyrolysis characteristics and bio-oil composition using the simplex lattice mixture design method. Energy Convers Manage 138:106–118
100. Zhang C, Hu X, Guo H, Wei T, Dong D, Hu G et al (2018) Pyrolysis of poplar, cellulose and lignin: effects of acidity and alkalinity of the metal oxide catalysts. J Anal Appl Pyrol 134:590–605
101. Pasangulapati V, Ramachandriya KD, Kumar A, Wilkins MR, Jones CL, Huhnke RL (2012) Effects of cellulose, hemicellulose and lignin on thermochemical conversion characteristics of the selected biomass. Biores Technol 114:663–669
102. Brillard A, Brilhac JF (2020) Improvements of global models for the determination of the kinetic parameters associated to the thermal degradation of lignocellulosic materials under low heating rates. Renew Energy 146:1498–1509
103. Nakashima K, Yamada L, Satou Y, Azuma JI, Satoh N (2004) The evolutionary origin of animal cellulose synthase. Dev Genes Evol 214(2):81–88
104. Matthysse AG, Deschet K, Williams M, Marry M, White AR, Smith WC (2004) A functional cellulose synthase from ascidian epidermis. Proc Natl Acad Sci 101(4):986–991
105. Bradbury AG, Sakai Y, Shafizadeh F (1979) A kinetic model for pyrolysis of cellulose. J Appl Polym Sci 23(11):3271–3280

106. Matsuoka S, Kawamoto H, Saka S (2014) What is active cellulose in pyrolysis? An approach based on reactivity of cellulose reducing end. J Anal Appl Pyrol 106:138–146
107. Matsuzawa Y, Ayabe M, Nishino J (2001) Acceleration of cellulose co-pyrolysis with polymer. Polym Degrad Stab 71(3):435–444
108. Shafizadeh F (1968) Pyrolysis and combustion of cellulosic materials. In: Advances in carbohydrate chemistry, vol 23, pp 419–474. Academic Press, Cambridge
109. Shafizadeh F (1984) The chemistry of pyrolysis and combustion
110. Paulsen AD, Mettler MS, Dauenhauer PJ (2013) The role of sample dimension and temperature in cellulose pyrolysis. Energy Fuels 27(4):2126–2134
111. Shen D, Hu J, Xiao R, Zhang H, Li S, Gu S (2013) Online evolved gas analysis by thermogravimetric-mass spectroscopy for thermal decomposition of biomass and its components under different atmospheres: part I. Lignin. Biores Technol 130:449–456
112. Shen D, Ye J, Xiao R, Zhang H (2013) TG-MS analysis for thermal decomposition of cellulose under different atmospheres. Carbohyd Polym 98(1):514–521
113. Mamleev V, Bourbigot S, Yvon J (2007) Kinetic analysis of the thermal decomposition of cellulose: the main step of mass loss. J Anal Appl Pyrol 80(1):151–165
114. Puskas JE (2013). Introduction to polymer chemistry: a biobased approach. DEStech Publications, Inc., USA
115. Dence CW (1992) The determination of lignin. In: Methods in lignin chemistry. Springer, Berlin, pp pp 33–61
116. Sharma S, Sharma A, Mulla SI, Pant D, Sharma T, Kumar A (2020) Lignin as potent industrial biopolymer: an introduction. In: Lignin, biosynthesis and transformation for industrial applications. Springer series on polymer and composite materials. Springer, Berlin, pp 1–15
117. Brebu M, Vasile C (2010) Thermal degradation of lignin—a review. Cellul Chem Technol 44(9):353
118. Ramiah MV (1970) Thermogravimetric and differential thermal analysis of cellulose, hemicellulose, and lignin. J Appl Polym Sci 14(5):1323–1337
119. Jiang G, Nowakowski DJ, Bridgwater AV (2010) Effect of the temperature on the composition of lignin pyrolysis products. Energy Fuels 24(8):4470–4475
120. Nassar MM, MacKay GDM (1984) Mechanism of thermal decomposition of lignin. Wood Fiber Sci 16(3):441–453
121. Rowell RM (2012) Handbook of wood chemistry and wood composites. CRC Press, Boca Raton
122. Müller-Hagedorn M, Bockhorn H, Krebs L, Müller U (2002) Investigation of thermal degradation of three wood species as initial step in combustion of biomass. Proc Combust Inst 29(1):399–406
123. Li DL, Ge SB, Peng WX, Wu QD, Wu JG (2017) Chemical structure characteristics of wood/lignin composites during mold pressing. Polym Compos 38(5):955–965
124. Yu J, Paterson N, Blamey J, Millan M (2017) Cellulose, xylan and lignin interactions during pyrolysis of lignocellulosic biomass. Fuel 191:140–149
125. Poletto M, Zattera AJ, Santana RM (2012) Thermal decomposition of wood: kinetics and degradation mechanisms. Biores Technol 126:7–12
126. Grønli MG, Várhegyi G, Di Blasi C (2002) Thermogravimetric analysis and devolatilization kinetics of wood. Ind Eng Chem Res 41(17):4201–4208
127. Maryandyshev P, Chernov A, Lyubov V, Trouvé G, Brillard A, Brilhac JF (2015) Investigation of thermal degradation of different wood-based biofuels of the northwest region of the Russian Federation. J Therm Anal Calorim 122(2):963–973
128. Lowden LA, Hull TR (2013) Flammability behaviour of wood and a review of the methods for its reduction. Fire Sci Rev 2(1):1–19
129. Zeng K, Gauthier D, Minh DP, Weiss-Hortala E, Nzihou A, Flamant G (2017) Characterization of solar fuels obtained from beech wood solar pyrolysis. Fuel 188:285–293
130. Shi L, Chew MYL (2013) Fire behaviors of polymers under autoignition conditions in a cone calorimeter. Fire Saf J 61:243–253

References

131. Kremer I, Tomić T, Katančić Z, Hrnjak-Murgić Z, Erceg M, Schneider DR (2021) Catalytic decomposition and kinetic study of mixed plastic waste. Clean Technol Envir 23(3):811-827
132. Friedrich K, Walter R (eds) (2020) Structure and properties of additive manufactured polymer components. Woodhead Publishing
133. Canetti M, Bertini F, De Chirico A, Audisio G (2006) Thermal degradation behaviour of isotactic polypropylene blended with lignin. Polym Degrad Stab 91(3):494–498
134. Bertini F, Canetti M, Audisio G, Costa G, Falqui L (2006) Characterization and thermal degradation of polypropylene–montmorillonite nanocomposites. Polym Degrad Stab 91(3):600–605

Chapter 2
The Correlation Between External Heat Flux and Time to Ignition

2.1 General Assumptions

Time to ignition is one of the fundamental characteristics used to describe the onset of combustion. Although it is theoretically possible to distinguish the initiation of reactions in the gaseous phase (homogeneous combustion) and the initiation of oxidising reactions on the interface of phases (heterogeneous combustion), only the first is used in practice. The time to ignition may be characterised as the time necessary to start flaming combustion under trial conditions.

When a flammable substance is exposed to a source of thermal radiation, the thermal energy that reaches its surface may, from the perspective of the onset of combustion, be divided into three components:

1. A component reflected back into the environment
2. A component that heats a thin layer on the surface of the material
3. A component that penetrates into the depths of the material.

With regard to the component reflected back into the environment, a material is characterised by the emissivity or reflectivity of its surface. Most solid flammable materials are too thick to have any significant transmittance during the time necessary for their ignition. We may therefore say that:

$$\rho_R = \frac{q_a}{q_e} = 1 - \varepsilon \tag{2.1}$$

where ρ_R is the reflectivity, q_a is the component of the external heat flux reflected back into the environment, q_e is the external heat flux that reaches the surface of the material and ε is the emissivity. Since emissivity is related to the radiation of heat, a more accurate term would be absorptivity (α_{abs}), which is the ratio of the radiation absorbed by the surface to the radiation reaching the surface:

$$\alpha_{abs} = \frac{q_e - q_a}{q_e} = \varepsilon \tag{2.2}$$

© The Author(s), under exclusive license to Springer Nature Switzerland AG 2022
P. Rantuch, *Ignition of Polymers*, Springer Series on Polymer and Composite Materials,
https://doi.org/10.1007/978-3-031-13082-3_2

Table 2.1 Average absorptivity of selected polymers

Material	$\bar{\alpha}$		Source
	Flame radiation	Solar radiation	
Gum rubber	0.89	0.69	[1]
Neoprene rubber	0.67	0.35	[1]
Chloroprene	0.71	0.62	[1]
Polyphenylene oxide	0.88	0.48	[1]
Polypropylene	0.86	0.62	[1]
Polyethylene, low density	0.93	0.57	[1]
Polyvinyl chloride (grey)	0.91	0.89	[1]
Polyvinyl chloride (clear 0.33 cm)	0.85	0.15	[1]
Silicone rubber	0.79	0.62	[1]
Nylon 6/6	0.93	0.62	[1]
Polystyrene (clear)	0.78	0.095	[1]
Phenolic bakelite	0.91	0.78	[1]
Cork	0.60	0.52	[1]
Ash wood	0.76	0.36	[2]
Balsa wood	0.75	0.35	[2]
Birch wood	0.77	0.39	[2]
Mahogany wood	0.76	0.52	[2]
Maple wood	0.76	0.44	[2]
Oak wood	0.77	0.49	[2]
Spruce wood	0.76	0.35	[2]
White pine wood	0.76	0.43	[2]
Masonite wood	0.75	0.61	[2]

The absorptivity of selected polymers is shown in Table 2.1.

The second component comprised the heat absorbed by the material that is eventually used to heat the surface layer. In this case, the characteristic feature of the material is its specific heat capacity:

$$c = \frac{q_b t}{\rho L_s \Delta T} \qquad (2.3)$$

c is the specific heat capacity, q_b is the component of external heat flux that heats its surface layer, t is the exposure time to the external heat flux, ρ is the density of the material, L_s is the thickness of the surface layer, and ΔT is the change in temperature of the surface layer.

As Table 2.2 clearly shows, the specific heat capacity depends on the temperature of the material. If the temperature range is restricted, the heat capacity of any phase of organic compounds may be represented adequately by an expression such as [3]:

2.1 General Assumptions

Table 2.2 The heat capacity of selected polymer materials (Speight et al. [3])

Material	T [K]	c_p [kJ kg^{-1} K^{-1}]
Poly(ethylene terephthalate)	300	1.172
	400	1.8203
	600	2.1136
Poly(propylene)	300	1.622
	300	2.099
	600	3.178
Poly(methyl methacrylate)	300	1.3755
	400	2.0766
	550	2.4323
Poly(oxymethylene)	300	1.283
	300	1.920
	600	2.292
Poly(styrene)	300	1.2230
	300	1.2730
	400	1.9322
	600	2.4417
Poly(acrylonitrile)	300	1.297
	370	1.624
Poly(tetrafluoroethylene)	300	0.9106
	300	1.028
	700	1.454
Poly(vinyl acetate)	300	1.183
	320	1.8409
	370	1.898
Poly(vinyl chloride)	300	0.9496
	360	1.457
	380	1.569
Poly(vinyl fluoride)	300	1.301

$$c = a_c + b_c T + c_c T^2 \quad (2.4)$$

a_c, b_c, and c_c are empirical constants, which may be calculated using the following formulae (Speight et al. [3]):

$$a_c = (c_{p1} - b_c T_1) - c_c T_1^2 \quad (2.5)$$

$$b_c = \frac{c_{p1} - c_{p2}}{T_1 - T_2} - [(T_1 + T_2)c_c] \qquad (2.6)$$

$$c_c = \frac{c_{p1}}{(T_1 - T_2)(T_1 - T_3)} + \frac{c_{p2}}{(T_2 - T_1)(T_2 - T_3)} + \frac{c_{p3}}{(T_3 - T_2)(T_3 - T_1)} \qquad (2.7)$$

c_{p1} is the specific heat capacity at T_1, c_{p2} is the specific heat capacity at T_2, and c_{p3} is the specific heat capacity at T_3.

For wood, the heat capacity may be calculated as [4]:

$$c = \frac{1.25T(1 + 0.025M)}{297} \qquad (2.8)$$

T is the temperature and M is the moisture.

The third component that reaches the surface of the sample is transferred by means of heat conduction into the cooler layers of the material. Heat conduction of materials is characterised by the thermal conductivity coefficient:

$$K = -\frac{q_c}{T_{grad}} \qquad (2.9)$$

K is the coefficient of thermal conductivity, q_c is the component of the external heat flux that reaches the deeper layers of the material, and T_{grad} is the temperature gradient. Thermal conductivity coefficients for selected polymers are shown in Table 2.3.

The thermal conductivity of amorphous polymers may be predicted based on the thermal conductivity at glass transition temperature. Bicerano [5] provides formulae for temperatures that exceed the glass transition temperature and for temperatures lower than the glass transition temperature.

If $T \leq T_g$, then:

$$K_T \approx K_{T_g} \left(\frac{T}{T_g}\right)^{0.22} \qquad (2.10)$$

If $T > T_g$:

$$K_T \approx K_{T_g} \left(1.2 - 0.2\frac{T}{T_g}\right) \qquad (2.11)$$

K_T is thermal conductivity at T, and K_{T_g} is thermal conductivity at glass transition temperature (T_g).

Tenwolde et al. [6] state that in the case of wood, the thermal conductivity may be determined based on its density in a dry state and its moisture content:

$$K = \left[\frac{\rho_d}{1000}(0.1941 + 0.004064M) + 0.01864\right] \qquad (2.12)$$

2.1 General Assumptions

Table 2.3 The thermal conductivity of selected polymer materials (Speight et al. [3])

Material		T [K]	K [W m^{-1} K^{-1}]
Poly(acrylonitrile butadiene styrene) copolymer	Injection moulding grade		0.33
Polyethylene	Low density		0.33
	Medium density		0.42
	High density		0.52
Poly(ethylene terephthalate)		20	0.15
Poly(methyl methacrylate)		20	0.21
Poly(oxymethylene)		20	0.292
Polypropylene		20	0.12
Polystyrene		0	0.105
		100	0.128
		200	0.13
		300	0.14
		400	0.16
	Foam (16 kg m^{-3})		0.040
	Foam (32 kg m^{-3})		0.036
	Foam (64 kg m^{-3})		0.033
	Foam (96 kg m^{-3})		0.036
	Foam (160 kg m^{-3})		0.039
Polytetrafluoroethylene		20	0.25
		72	0.34
Polyurethane	Casting resin	20	0.21
	Elastomer	20	0.31
Poly(vinyl chloride)	Rigid	20	0.21
	Flexible	20	0.17
	Chlorinated	20	0.14
	Foam (56 kg m^{-3})		0.035
	Foam (112 kg m^{-3})		0.040
Papers		303–333	0.029–0.17
Maple wood	Parallel to face	20	0.425
	Perpendicular to face	50	0.182
Oak wood	Parallel to face	15	0.349
	Perpendicular to face	15	0.209
Pine wood	Parallel to face	20	0.349
	Perpendicular to face	15	0.151

This correlation may be further modified for various temperatures [7]:

$$K = \frac{\left[\frac{\rho_d}{1000}(0.1941 + 0.004064M) + 0.01864\right]T}{297} \quad (2.13)$$

ρ_d is the volume weight of dry wood. However, it must be noted that this calculation is particularly suitable for wood with a volume weight over 300 kg m^{-3} and moisture lower than 25%. Moreover, the authors did not take the impact of temperature into consideration due to an insufficient amount of data. That being said, they suggest that the thermal conductivity of wood rises with increasing temperature. For non-solid wooden materials, for example, plywoods and OSB boards, the following formula applies [6]:

$$K_{\text{Plywood}} = 0.86 \times K_{\text{Wood}} \quad (2.14)$$

After the surface layer of the material heats up, a thermal gradient forms between the surface layer and the environment. Some of the heat used to heat up the top layer of the material is released into the environment in the form of surface heat loss. This includes heat loss by convection and by re-radiation.

$$q_{\text{loss}} = q_{\text{loss\_r}} + q_{\text{loss\_c}} = \varepsilon\sigma\left(T_s^4 - T_0^4\right) + h_c(T_s - T_0) \quad (2.15)$$

q_{loss} is the surface heat loss, $q_{\text{loss\_r}}$ is the heat loss through re-radiation, $q_{\text{loss\_c}}$ is the heat loss by convection, σ is Stefan–Boltzmann constant, T_s is the surface temperature, T_0 is the ambient temperature, and h_c is the convective heat transfer coefficient.

A schematic depiction of heat fluxes for thermally thick materials is illustrated in Fig. 2.1.

It must be noted that when the heat fluxes necessary for ignition are applied to the material, physical and chemical changes tend to occur that result, inter alia, in changes of emissivity, thermal capacity, and the thermal conductivity coefficient.

To initiate flame combustion, a flammable material has to release a sufficient quantity of flammable gases and steams. These will subsequently mix with the oxygen in the air and form a flammable mixture that may be ignited. The quantity of flammable gases released greatly depends on the temperature of the material, especially its surface layer. For this reason, external heat flux has a considerable impact on the time to ignition. This impact is clearly illustrated in Fig. 2.2.

2.1 General Assumptions

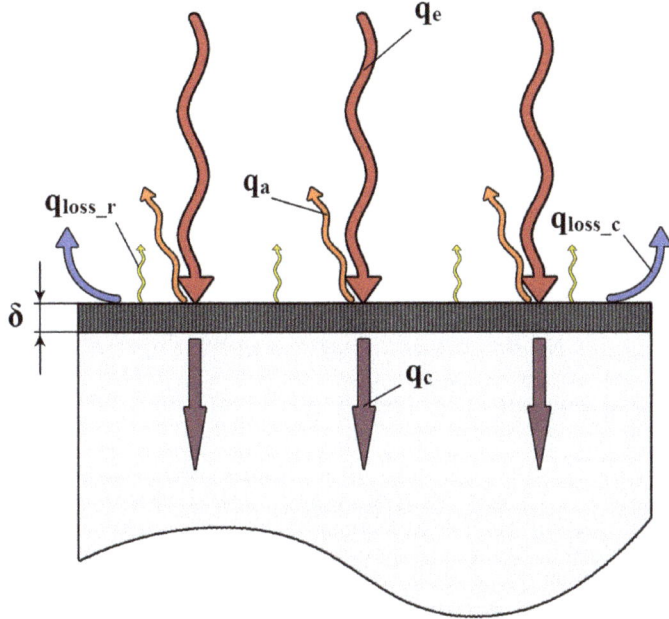

Fig. 2.1 A schematic depiction of heat transfer in a thermally thick material loaded by an external heat flux

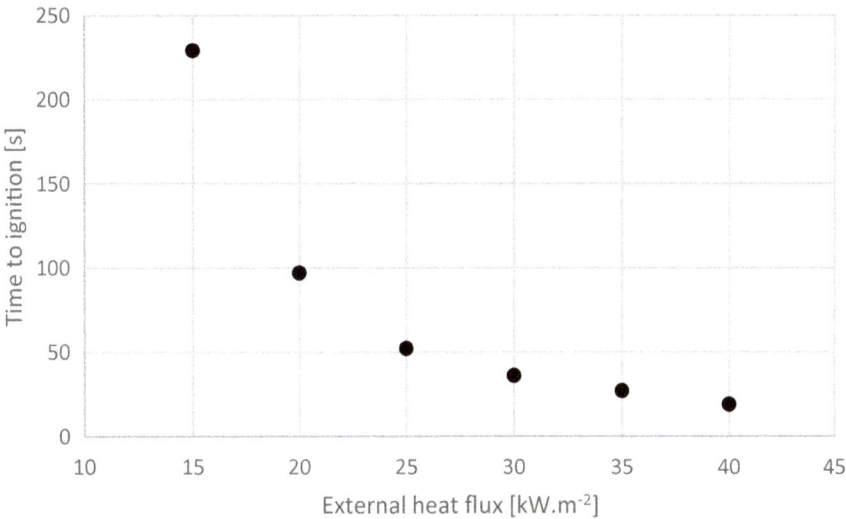

Fig. 2.2 An example of the correlation between the time to ignition and the external heat flux

2.2 Suggested Correlations

2.2.1 Determining Correlation According to Lawson and Simms [8]

Provided that:

1. The material is inert;
2. The surface cooling is Newtonian;
3. There is no appreciable rise in temperature of the back of the material during the experiment.

The following applies:

$$\Delta T = \frac{q}{\Psi}\left[1 - \frac{1}{\Psi}\left(\frac{K\rho c}{\pi t}\right)^{\frac{1}{2}}\right] \quad (2.16)$$

ΔT is the temperature rise, q is the radiation intensity, Ψ is the rate of loss of heat per unit area for each degree rise in temperature, K is the thermal conductivity, ρ is the density, c is the specific heat of the material, π is Ludolf's number, and t is time. If the sample is ignited when its surface reaches a sufficiently high temperature, the time to ignition depends on the intensity of radiation reaching its surface. Since each material only ignites when irradiated by an intensity greater than the critical intensity q_{cr}, the time to ignition when irradiated by an intensity q might be expected to depend upon the quantity $(q - q_{cr})$. Based on the formula log $(q - q_{cr})$ versus log t, it was discovered that [8]:

$$(q - q_{cr})t_{ig}^{\frac{2}{3}} = A \quad (2.17)$$

$$(q - q_{cr})t_{ig}^{\frac{4}{5}} = B \quad (2.18)$$

A is the constant for pilot ignition, and B is the constant for autoignition. Constants A and B vary for the different species investigated.

Based on the measured data (Table 2.4), it is possible to derive formulae for pilot ignition and spontaneous ignition [8]:

$$A = 0.25 \times 10^6 (K\rho c + 68 \times 10^{-6}) \quad (2.19)$$

$$B = 0.05 \times 10^6 (K\rho c + 35 \times 10^{-6}) \quad (2.20)$$

Then, for piloted ignition:

2.2 Suggested Correlations

Table 2.4 Results of the measurements from [8]

Material	L_0 [mm]	q_{cr} for spontaneous ignition [kW m^{-2}]	q_{cr} for pilot ignition [kW m^{-2}]
Fibre insulation board	12.7	23.86	6.28
Western red cedar	19.05	26.80	14.65
American whitewood	19.05	25.54	14.65
Freijo	19.05	26.38	15.07
African mahogany	19.5	23.86	12.56
Oak	19.5	27.63	15.07
Iroko	19.5	–	15.07

$$t_{ig} = \left(\frac{0.25 \times 10^6 (K\rho c + 68 \times 10^{-6})}{(q - q_{cr})} \right)^{\frac{3}{2}} \quad (2.21)$$

And for autoignition:

$$t_{ig} = \left(\frac{0.05 \times 10^6 (K\rho c + 35 \times 10^{-6})}{(q - q_{cr})} \right)^{\frac{5}{4}} \quad (2.22)$$

It should be noted that in this case, heat flux was calculated in cal cm^{-2} s^{-1}, density was determined in g cm^{-3} and specific heat in cal g^{-1} °C^{-1}. Calorie (cal) is a historical unit of heat. It is equal to about 4.18 J.

In the experimental work, the authors used a surface combustion heater with a surface area of 1 ft^2 (approx. 929 cm^2) as the source of external heat flux. A mixture of coal gas/air was used as fuel. The heat fluxes used ranged from 0.15 cal cm^{-2} s^{-1} (approx. 6.28 kW m^{-2}) to 1.5 cal cm^{-2} s^{-1} (approx. 62.76 kW m^{-2}). The samples of wood were oven-dried, and the surface area exposed to external heat flux was 2 in.^{2} (approx. 12.9 cm^2). The ignitor was a flame held 0.5 in. (approx. 1.27 cm) above and in front of the surface to be irradiated [8].

2.2.2 Determining Correlation According to Koohyar [9], Hallman [1] and Wesson et al. [2]

Koohyar observed the ignition of wood caused by the radiation of flames. He indicates that this process can be described by a mathematical model for an inert and opaque solid with the surface boundary condition [9]:

$$-KT_{\text{grad}} = n_K q_e \tag{2.23}$$

T_{grad} is the temperature gradient at the surface, and n_K is a constant. In the experimental part of his work, he used samples of five wood species that were thermally exposed to a flame in a vertical direction. Measurements were conducted in the vertical position both with a flame ignitor situated above the sample and without it. The thickness of the samples was 0.5 in. (1.27 cm), 0.625 in. (1.59 cm), and 0.75 in. (1.91 cm), respectively. The exposed area of the samples was a 3.9 in. (9.91 cm) square [9].

In his dissertation thesis, Wesson sets out the following correlation for wood:

$$t_{\text{ig}} = 80 \left[\frac{\rho^{\frac{1}{3}}}{q_i} \right] \tag{2.24}$$

It should be noted that in this case, incident heat flux (q_i) was calculated in cal cm^{-2} s^{-1}, and density was determined in g cm^{-3}. The measurements were conducted in the vertical position on three wood species (Alaska cedar, cottonwood, and oak). Prior to testing, the samples of wood were oven-dried for at least 24 h. They were 1.97 cm and 2.54 cm thick, respectively. The nominal exposed area of the samples was 9.9 cm × 9.9 cm. The device used for these measurements was designed so that samples could be loaded by radiation heat from open diffuse flames as well as from tungsten lamps. A heated wire coil acting as an ignitor was placed above the samples [10].

Hallmann built upon the measurements of Koohyar and Wesson. He observed the process of the ignition of plastics and rubber, and the techniques used in his dissertation were essentially those previously used by Koohyar or Wesson. Measurements were conducted on samples of 33 polymers including plexiglas, poly(vinyl chloride), buna rubbers, phenolics, nylon, polyethylene, polypropylene, polyurethane, cellulosic materials, neoprene rubbers, butyl rubber, and silicone rubber. The samples had a thickness of 1.27 cm, and the exposed surface was 8.8 cm × 10.9 cm [1]. The starting point for the determination of the correlation was the general model provided by Carslaw and Jaeger [11]. Hallman [1] presents base equation as:

$$\Delta T_s = \frac{\overline{\alpha} q_e t^{\frac{1}{2}}}{(K\rho c)^{\frac{1}{2}}} \sum_{n=0}^{\infty} \left[\text{ierfc} \frac{2nL}{2(\kappa t)^{\frac{1}{2}}} + \text{ierfc} \frac{(2n+2)L}{2(\kappa t)^{\frac{1}{2}}} \right] \tag{2.25}$$

ierfc denotes the first integral of the complementary error function, κ is thermal diffusivity, and t is time.

Firstly, based on the general model and data from literature, the author determined the relation [1]:

$$t_{\text{ig}} = \Phi \frac{(T_{\text{ig}} - T_0)^2 (K\rho c)}{\overline{\alpha} q_e} \tag{2.26}$$

2.2 Suggested Correlations

Φ is a constant, and $\overline{\alpha}$ is average absorptivity.

According to formula (2.26), time to ignition should be a function of the difference between the temperature of the surface of the material, the thermal inertia, and the heat flux absorbed by the material. After applying the measured data and using the method of least squares, Hallman [1] suggested the following formula for plastics [1]:

$$t_{ig} = 160 \frac{(T_{ig} - T_0)^{1.04} (K\rho c)^{\frac{3}{4}}}{(\overline{\alpha} q_e)^2} \tag{2.27}$$

Wesson et al. [2] worked with the same general model as Hallman, but they applied it to 13 species of wood. After adjustments, they achieved the following formula:

$$t_{ig} = 35 \frac{\rho^{0.9} \left\{ \mathrm{erf}\left[\frac{L}{2}(\kappa t_{ig})^{\frac{1}{2}}\right]\right\}^{1.2}}{(\overline{\alpha} q_e)^{2.8}} \tag{2.28}$$

erf is the error function.

It should be noted that in both cases, heat flux was calculated in cal cm^{-2} s^{-1}, density was determined in g cm^{-3} and specific heat in cal g^{-1} °C^{-1}. The samples were dried prior to taking measurements. Their thickness ranged from less than 0.2 cm to more than 2.5 cm. Similar to previous studies, the sources of heat were diffused flames and high-temperature tungsten lamps, with a heat flux that ranged from 0.6 cal cm^{-2} s^{-1} (25.1 kW m^{-2}) to 3.5 cal cm^{-2} s^{-1} (146.44 kW m^{-2}) [2].

2.2.3 Determining Correlation According to Smith and Satija [12]

In their study, Smith and Satija [12] dealt with a mathematical model to predict fire growth in a compartment. They state that the surface temperature is not an adequate measure of "ignition" point for many products. They suggest using the flux–time product to describe ignition, which may be defined by the following formula:

$$\mathrm{FTP} = \sum \left[(\mathrm{Flux} - \mathrm{SPF})_{\Delta t}^n \Delta t \right] \tag{2.29}$$

FTP is the flux–time product, Flux is the average incident flux over time increment, SPF is the self-propagation flux, Δt is the time span of incremented time, and n is the empirical constant [12].

Formula 2.29 for time to ignition may be written as follows:

$$t_{ig} = \frac{(\mathrm{FTP})}{(q_e - q_{cr})^n} \tag{2.30}$$

This method was eventually used to describe the correlation between time to ignition and external heat flux when using a conical heat source [13].

2.2.4 Determining Correlation According to Quintiere and Harkleroad [14]

To describe ignition based on the measured data, they used the following formula [14]:

$$\frac{q_{cr}}{q} = b\sqrt{t} \tag{2.31}$$

b is a constant for each material. At the same time, they suggest the following relationship for time to ignition and thermally thick materials [14]:

$$K\rho c = \frac{4}{\pi}\left(\frac{h}{b}\right)^2 \tag{2.32}$$

h is the heat loss coefficient.

By combining formulae (2.31) and (2.32), we obtain the following:

$$t = \pi K\rho c \left(\frac{q_{cr}}{2q_e h}\right)^2 \tag{2.33}$$

These formulae were correlated using measurements from a gas-fired radiant panel and a framed sample-holder assembly that contained a vertically oriented sample. The device allows the observation of ignition and the propagation of flame on the surface of a sample. The ignitor of combustion was an acetylene-air pilot flame located approximately 25 mm from the top of the sample. The samples tested were a wide range of solid flammable substances. Before the measurements, they were conditioned at a temperature of 20–22 °C and humidity of 55%. They measured 162 mm in height and 806 mm in length. The first 110 mm of the length of the sample was used for the observation. During the measurements, they were supported in the spring-loaded frame by a calcium silicate board 13 mm thick [14]. This method was also used by Hu and Clark [15] for measurements using a cone heater.

2.2.5 Determining Correlation According to Bluhme [16]

Bluhme [16] based his theories on the assumption that in general, the correlation between time to ignition and external heat flux may be described as:

2.2 Suggested Correlations

$$t_{ig} = \left(\frac{q_e}{a}\right)^{\frac{1}{z}} \tag{2.34}$$

a and z are constants. Based on the measured data, he proposed specific formulae for plasterboard and wood, specifying the range in which the formulae may be applied. In the case of plasterboard, the relationship is as follows:

$$t_{ig} = \frac{1}{(1.11 q_e - 27.8) \times 10^{-3}} \tag{2.35}$$

$$30 \,\text{kW m}^{-2} \leq q_e \leq 40 \,\text{kW m}^{-2} \tag{2.36}$$

For wood, Bluhme determined the following correlation:

$$t_{ig} = \frac{1}{(1.11 q_e - 16.7) \times 10^{-3}} \tag{2.37}$$

$$20 \,\text{kW m}^{-2} \leq q_e \leq 30 \,\text{kW m}^{-2} \tag{2.38}$$

The results were obtained through measurements conducted according to ISO 5657:1986 [17] with external heat fluxes of 20 kW m^{-2}, 30 kW m^{-2}, 40 kW m^{-2}, and 50 kW m^{-2} [16]. This method uses square-shaped samples of 165 mm × 165 mm with thickness of up to 70 mm. A circular surface area of the sample with a diameter of 140 mm is exposed to the external heat flux. A dried, non-flammable, insulating board of 6 mm in thickness serves as a baseboard. The external heat flux is provided by a cone heat source with an upper diameter of 66 mm and a lower diameter of 200 mm. The testing takes place horizontally. A pilot flame provides the ignition source for flaming combustion, which is moved into its test position (10 mm above the centre of the sample) every 4 s [17].

2.2.6 Determining Correlation According to Mikkola and Wichman [18]

Mikkola and Wichman [18] correlated the ignition data of various materials with different thermophysical properties and overall thermal behaviours. By doing so, they created a simplified model, in which the following aspects were ignored for the sake of simplicity.

- Gas phase complications such as the interaction of incident radiation with volatile gases leaving the surface and the influences of gas motion along the surface
- Solid phase complications, such as variable radiant absorption thickness, variable thermophysical and thermochemical properties, and heat and mass transfer interactions in the decomposing solid phase.

The Linearised Thermal Ignition Model

In this case, the heat losses from the surface of the sample are linearised. Thus, the radiant losses are of the same functional form as the convective heat losses, so the overall surface heat loss coefficient is the sum of convective and radiative parts [18].

The specified correlation for a thermally thin material is following:

$$t_{ig} = \rho c L_0 \frac{(T_{ig} - T_0)}{q_o} \qquad (2.39)$$

L_0 is sample thickness and q_o is overall heat flux, and it also includes surface heat losses.

For a thermally thick material, the following formula applies:

$$t_{ig} = \frac{\pi}{4} K\rho c \left(\frac{T_{ig} - T_0}{q_n} \right)^2 \qquad (2.40)$$

q_n is net heat flux.

The General Integral Model

The authors based their work on the exact integral energy balance of a simple thermal model, which they reduce to:

$$q_{in} - q_{out} = \rho c \frac{\partial}{\partial t} \int_0^{L_0} (T - T_0) dy \qquad (2.41)$$

q_{in} is the sum of total heat fluxes entering the sample, q_{out} is the sum of total heat fluxes leaving the sample, t is time, T is temperature, and y is the spatial co-ordinate. Subsequently, they derived formulae for the individual materials based on their thermal thickness.

For thermally thin materials:

$$t_{ig} \approx \rho c L_0 \frac{T_{ig} - T_0}{q_{in} - q_{out}} \qquad (2.42)$$

For thermally intermediate thickness materials:

$$t_{ig} \sim \rho c \sqrt{K L_0} \left(\frac{T_{ig} - T_0}{q_{in} - q_{out}} \right)^{\frac{3}{2}} \qquad (2.43)$$

For thermally thick materials:

$$t_{ig} \approx K\rho c \left(\frac{T_{ig} - T_0}{q_{in} - q_{out}} \right)^2 \qquad (2.44)$$

2.2 Suggested Correlations

The authors used results from various other authors to validate their models. At first, this included data from the initiation of combustion of various wood species, then they focussed on the time to ignition of woods obtained from measurements using a cone calorimeter, and finally, they shifted their focus to particle board, plywood, polyurethane, poly(vinylchloride), polypropylene, and polymethylmethacrylate. Based on these comparisons, Mikkola and Wichman [18] state that the results for most samples of natural wood correlate well with the thermally simple formulae; however, certain wood-based materials (such as plywood) and certain plastics (such as PMMA) do not behave according to these models. They point out various conditions that affect the measurements:

1. The surface smoothness must be as nearly uniform as possible because roughening alters the heat flux absorption area;
2. The data for large external heat fluxes (which produce short times to ignition, often less than 10 s) is usually inaccurate;
3. Ideally, the experimental apparatus should not alter the thermal thickness of the sample;
4. The data for very low heat fluxes is very sensitive to minor changes in both external and internal conditions;
5. All data is gathered for radiative external heating and is correlated as though the experimental configurations were always horizontal.

It is apparent that the authors' work is based on the so-called thermal thickness of the material. Since its thermal thickness, apart from the characteristics typical of the material under test, also depends on its physical thickness, we cannot speak of a characteristic of the material, but rather of the specific sample.

Thermally thin materials are those with a rate of thermal conductivity into the material that is significantly higher than the rate at which the surface temperature increases (Fig. 2.3). This state occurs in physically thin materials or materials with extremely high thermal conductivity. Examples of thermally thin materials include textiles, individual sheets of paper, or metal sheets. Most materials with a thickness greater than 1 mm behave like thermally thick materials in a fire. There is a significant thermal gradient between the side exposed to fire and the unexposed side. The heat transfer from the unexposed surface has no substantial impact on the heat transfer into the exposed surface (Almirall and Furton [26]).

However, the thermal thickness does not necessarily have to be the same as the physical thickness, and materials with a final thickness that behave as thermally thick objects during initial heating will react as thermally thin materials after a sufficient amount of time [19].

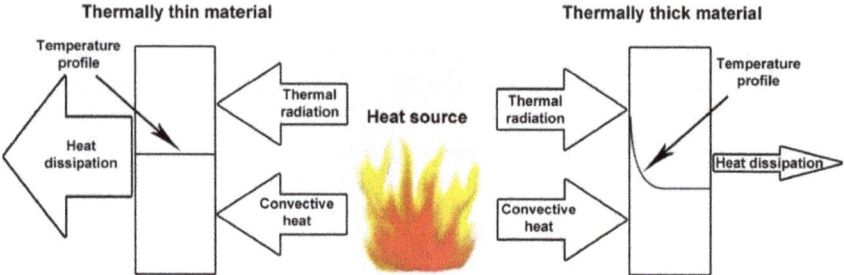

Fig. 2.3 Heat load comparison of a thermally thin and a thermally thick material

2.2.7 Determining Correlation According to Delichatsios et al. [20]

The authors used a single-dimensional heat transfer equation to model heat propagation into a solid substance. The following simplifications were used [20]:

- The material was considered to be thermally thick;
- Convection losses were ignored;
- The thermal properties of materials were considered independent of temperature;
- The imposed heat flux was constant.

One of the possible formulae for this correlation is:

$$t_{ig} = \frac{\pi}{4} K\rho c \left(\frac{T_p - T_0}{q - 0.64\sigma\left(T_p^4 - T_0^4\right)} \right)^2 \quad (2.45)$$

q is the applied heat flux, σ is the Stefan–Boltzmann radiation constant, and T_p is the surface temperature of the material at ignition.

Formula 2.45 is a good approximation for:

$$\frac{T_p - T_0}{T_{max} - T_0} < 0.7 \quad (2.46)$$

$$\frac{T_0}{T_{max}} < 0.4 \quad (2.47)$$

T_{max} is the maximum surface temperature that could be obtained from a non-pyrolysing material.

If

$$\frac{q}{q_{cr}} < 1.1 \quad (2.48)$$

then

2.2 Suggested Correlations

$$t_{ig} = \frac{K\rho c}{\pi} \left(\frac{T_p - T_0}{q - q_{cr}}\right)^2 \qquad (2.49)$$

The authors obtained actual values to be compared with their theoretical model through measurements made using a flammability apparatus. This apparatus uses inclined quartz heaters to heat up the samples, with the quartz heaters and the samples placed in a pyrolysis chamber with an air supply from the bottom. The samples were made of polymethylmethacrylates painted with a layer of carbon black [20].

2.2.8 Determining Correlation According to Janssens [7]

This model was designed for wood-based materials. It is based on the following assumptions [7]:

- The heat flux of solid materials is considered to be single-dimensional;
- Any chemical change in the material prior to ignition is negligible;
- Heat transmission by conductivity between volatile substances and the surface of the material is negligible;
- Ignition occurs at the moment that a sufficiently high surface temperature is reached;
- The material is non-transparent;
- Absorptivity equals emissivity and along with reflectivity they add up to 1;
- Absorptivity, emissivity, and reflectivity are constant;
- The heat losses from the surface are partly radiative and partly convective with a constant convection coefficient;
- The specimens behave as semi-infinite solids.

As a basis for the derivation of the correlation between critical heat flux and the time to ignition, they used the following formula:

$$(q_e - q_{cr})t_{ig}^n = \text{Constant} \qquad (2.50)$$

If non-dimensional time is defined as:

$$\tau \equiv \frac{h_{ig}^2 t}{K\rho c} \qquad (2.51)$$

τ is the non-dimensional time and h_{ig} is the convection coefficient at ignition:

$$h_{ig} \equiv \frac{\alpha_{abs} \cdot q_{cr}}{T_{ig} - T_0} \qquad (2.52)$$

Then, for values of non-dimensional time lower than 20 (corresponding to 24 min for wood), and taking linearised heat losses into consideration, they obtained the

Table 2.5 Wood samples used in the work of Janssens [7]

Wood type	ρ_d [kg m^{-3}]	L_0 [mm]
Softwoods		
Western cedar	330	17.0
Redwood	430	19.0
Hardwoods		
Radiata pine	460	17.5
Douglas fir	465	16.8
Victorian ash	640	17.2
Blackbutt	810	17.4

following correlation:

$$F(\tau) \approx \frac{1}{1 + 0.73\tau^{-0.547}} \qquad (2.53)$$

Their calculations subsequently proved that this correlation also applies to non-linear surface heat losses as well as to the temperature-dependent thermal properties of wood. Their final formula may be adjusted to the following form:

$$t_{ig} = \frac{0.5625 K \rho c}{h_{ig}^2 \left(\frac{q_e}{q_{cr}} - 1\right)^{1.828}} \qquad (2.54)$$

Experimental data, obtained from the piloted ignition of six oven-dried wood species (Table 2.5) using a cone calorimeter, was used for the determination of a correlation. External heat fluxes from 15 to 45 kW m^{-2} were used. The samples were square-shaped, with 10 cm sides [7].

2.2.9 Determining Correlation According to Spearpoint and Quintiere [21]

Spearpoint and Quintiere [21] proposed an integral model based on the following presumptions:

- Ignition occurs at the moment when a critical surface temperature is reached;
- The material is inert until the moment of ignition;
- The sample is infinitely thick.

The authors derived the following formula:

2.2 Suggested Correlations

Table 2.6 Wood samples used in the work of Spearpoint and Quintiere [21]

Wood type	Grain orientation	M [%]
Douglas fir	Along	8.6
	Across	7.4
Redwood	Along	5.1
	Across	5.2
Red oak	Along	7.4
	Across	8.5
Maple	Along	4.8
	Across	4.8

$$t_{ig} = \frac{4}{3}\left[\frac{1}{(2-\beta_{ig})(1-\beta_{ig})}\right]K\rho c \frac{(T_{ig}-T_0)^2}{q_i^2} \quad (2.55)$$

β_{ig} is the ratio of convective gain and radiative loss with incident heat flux at ignition.

Wood samples (Table 2.6) were measured with the grain parallel to the incident heat flux and perpendicular to the incident heat flux. They were 50 mm thick, and the surface area exposed to heat flux was 96 mm². Prior to taking measurements, they were stored at 50% humidity and a temperature of 20 °C. The measurements were conducted in a horizontal orientation using a cone calorimeter [21].

2.2.10 Determining Correlation According to Harada [22]

To determine a correlation for the calculation of time to ignition of wood exposed to an external heat flux, Harada [22] made the following assumptions:

- Time to ignition depends on the external heat flux and thermal inertia;
- The thermal conductivity coefficient of wood has a linear correlation with density, through the application of the following formulae obtained by Urakami and Fukuyama from experimental data (Urakami and Fukuyama [27]) [22]:

$$K_\perp = (174 + 1.86\rho) \times 10^{-7} \quad (2.56)$$

$$K_\parallel = (232 + 4.02\rho) \times 10^{-7} \quad (2.57)$$

$K_\perp$ is thermal conductivity in the perpendicular direction, and $K_\parallel$ is thermal conductivity in the parallel direction;

- The specific thermal capacity is practically independent of the species of wood, and its value is 1.25 kJ kg^{-1} K^{-1}.

Based on his measurements of time to ignition, the external heat flux, and density, he determined the following correlation [22]:

$$t_{ig} = 14.4 \frac{K\rho c}{q_e^3} + 8.64 \tag{2.58}$$

Harada [22] conducted measurements on nine wood species with thicknesses that ranged from 10 to 40 mm (Table 2.7). Square samples of wood with a length of 100 mm were prepared for testing under a heat flux, and they were radial, tangential, or cross-sectional samples. Before the tests, the samples were dried for more than 48 h at a temperature of 60 °C, and the drying process was completed using silica gel in a desiccator at room temperature. The final moisture content of the samples was less than 3%. A cone calorimeter was used as the test apparatus, and heat fluxes of 20 kW m^{-2}, 25 kW m^{-2}, 30 kW m^{-2}, 40 kW m^{-2}, and 50 kW m^{-2} were applied to the samples.

Table 2.7 Wood samples used by Harada [22]

Wood type	ρ [kg m^{-3}]	Vessel elements [%]	L_0 [mm]
Softwoods			
Japan cedar	299	Not stated by the author	10; 20; 40
Hiba arborvitae	422	Not stated by the author	10; 20; 40
Japanese red pine	433	Not stated by the author	10; 20; 40
Japanese larch	542	Not stated by the author	10; 20; 40
Hardwoods			
Paulownia	266	14.8	20
Japanese walnut	547	17.2	10; 20; 40
Japanese beech	581	41.7	10; 20; 40
Zelkova	703	18.2	20
Japanese oak	772	10.7	10; 20; 40

2.2.11 Determining Correlation According to Shi and Chew [23]

Shi and Chew [23] worked on the determination of the correlation between the time to ignition and the external heat flux at the autoignition of wood. After a review of the works of other authors, they proposed five general formulae:

$$t_{ig} = a_1 K\rho c \left(\frac{T_{ig} - T_0}{q_e}\right)^2 \tag{2.59}$$

$$t_{ig} = a_2 K\rho c \left(\frac{T_{ig} - T_0}{q_e - 28.0}\right)^2 \tag{2.60}$$

$$t_{ig} = a_3 K\rho c \left(\frac{T_{ig} - T_0}{q_e - 18.0}\right)^2 \tag{2.61}$$

$$t_{ig} = a_4 \frac{K\rho c}{q_e^3} + b_4 \tag{2.62}$$

$$t_{ig} = a_5 \frac{\rho^{0.73}}{(q_e - 28.0)^{1.82}} \tag{2.63}$$

a_1, a_2, a_3, a_4, a_5, and b_4 are constants for the individual formulae. Based on the measured data, formulae 1, 2, 3, and 5 gave R^2 values that exceeded 0.9. Due to the difficulty of obtaining data regarding ignition temperature, the authors prefer Formula (2.63). After calculating the constants, they suggested the following formula for the autoignition of wood:

$$t_{ig} = \frac{144\rho^{0.73}}{(q_e - 28.0)^{1.82}} \tag{2.64}$$

They tested samples of various species of wood (Table 2.8) that measured 100 mm × 100 mm, with a thickness of 10 mm, 20 mm, and 30 mm. They were placed horizontally, and a cone calorimeter was used as the testing apparatus. External heat fluxes of 25 kW m^{-2}, 50 kW m^{-2}, and 75 kW m^{-2} were applied [23].

2.2.12 Determining Correlation According to Babrauskas [24]

Babrauskas [24] suggests that important factors that affect the time to ignition of wood include density, thermal conductivity, and moisture, with the possibility of the determination of thermal conductivity based on density. He used 245 points to

Table 2.8 Wood samples used in the work of Shi and Chew [23]

Wood type	ρ [kg m^{-3}]	M [%]
Softwoods		
Pine	446.8	12.1
Hardwoods		
Beech	632.4	10.4
Cherry	557.6	11.3
Oak	897.3	10.4
Maple	748.2	10.0
Ash	680.0	10.2

postulate the correlation. He notes that the effect of density was identified statistically, but that the effect of moisture and orientation were swamped by data scatter. He proposed a correlation with the following formula:

$$t_{ig} = \frac{130\rho^{0.73}}{(q_e - 11.0)^{1.82}} \quad (2.65)$$

To reduce the impact of various testing methods, Babrauskas only used test data using a cone calorimeter. The root-mean-square error of the predictions was 64%, which indicates that predicting times to ignition can only be done semi-quantitatively, but this must also be placed in the context that the experimental data varied from 2.5 to 4200 s. Below about 15 kW m^{-2}, the points systematically deviate above a straight line. This might be expected, since the theory is based on a thermally thick material, and wood specimens that are 12–25 mm thick cease to behave in a thermally thick manner once heated for a long time [24].

2.2.13 Determining Correlation According to An et al. [25]

When determining a correlation between the time to ignition and the external heat flux applied to the surface of flammable materials, it is commonly presumed that the distance of the sample from the heater until the moment of ignition is constant. For most materials, this assumption is almost completely true. In the case of foam plastics, this may result in significant deviations. These materials may contain a relatively low amount of polymer as well as a substantial amount of gas. While under a thermal load, the polymer melts and the space filled with gases is destroyed. The gas escapes into the environment, and the volume of the material decreases. As a result, the distance from the heat source increases, and the heat flux exposure decreases.

This phenomenon was discussed in the work of An et al. [25]. They based their work on the following assumptions:

- The melting time of the polymer is negligible compared to its time to ignition;
- The thickness of the melted polymer is negligible;
- For shorter distances, the decrease in the effect of the external heat flux is practically linear as the distance of the heat source increases.

Based on the above assumptions, they determined the following correlation [25]:

$$t_{ig} \propto \left[\frac{1}{(1.0142 - 0.0674L_0)q_e} \right]^2 \tag{2.66}$$

The measurements were conducted using a cone calorimeter with a spark ignitor. During the tests, the samples were placed horizontally and exposed to external heat fluxes of 25 kW m^{-2}, 35 kW m^{-2}, and 45 kW m^{-2}. Two types of foam polystyrene were used (expanded polystyrene and extruded polystyrene) with thicknesses of 2 cm, 3 cm, 4 cm, and 5 cm, respectively. The samples were 100 mm × 100 mm squares [25].

References

1. Hallman JR (1971) Ignition characteristics of plastics and rubber. The University of Oklahoma, Oklahoma
2. Wesson HR, Welker JR, Sliepcevich CM (1971) The piloted ignition of wood by thermal radiation. Combust Flame 16(3):303–310
3. Speight J (2005) Lange's handbook of chemistry. McGraw-Hill Education, New York
4. Dietenberger MA (1996) Ignitability analysis using the cone calorimeter and LIFT apparatus. In: Proceedings of the international conference on fire safety: July 22–26, 1996, Columbus, OH, vol 22. Product Safety Corporation, pp 189–197
5. Bicerano J (2002) Prediction of polymer properties, 3rd edn. CRC Press, New York
6. TenWolde A, McNatt JD, Krahn L (1988) Thermal properties of wood and wood panel products for use in buildings (No. DOE/OR/21697-1). Forest Service, Madison, WI (USA). Forest Products Lab
7. Janssens M (1991) Fundamental thermophysical characteristics of wood and their role in enclosure fire growth. Doctoral dissertation, Ghent University
8. Lawson DI, Simms UD (1952) The ignition of wood by radiation. Br J Appl Phys 3(9):288
9. Koohyar AN (1968) Ignition of wood by flame radiation. The University of Oklahoma, Oklahoma
10. Wesson HR (1970) The piloted ignition of wood by radiant heat. The University of Oklahoma, Oklahoma
11. Carslaw HS, Jaeger JC (1959) Conduction of heat in solids (No. 536.23). Clarendon Press
12. Smith EE, Satija S (1983) Release rate model for developing fires
13. Toal BR, Silcock GWH, Shields TJ (1989) An examination of piloted ignition characteristics of cellulosic materials using the ISO ignitability test. Fire Mater 14(3):97–106
14. Quintiere JG, Harkleroad MT (1985) New concepts for measuring flame spread properties. In: Fire safety: science and engineering. ASTM International, USA
15. Hu X, Clark FR (1988) The use of the ISO/TC92 test for ignitability assessment. Fire Mater 12(1):1–5

16. Bluhme DA (1987) ISO ignitability test and proposed criteria. Fire Mater 11(4):195–199
17. ISO 5657:1986. Reaction to fire tests—ignitability of building products using a radiant heat source
18. Mikkola E, Wichman IS (1989) On the thermal ignition of combustible materials. Fire Mater 14(3):87–96
19. Babrauskas V (2002) Ignition of wood: a review of the state of the art. J Fire Prot Eng 12(3):163–189
20. Delichatsios MA, Panagiotou TH, Kiley F (1991) The use of time to ignition data for characterizing the thermal inertia and the minimum (critical) heat flux for ignition or pyrolysis. Combust Flame 84(3–4):323–332
21. Spearpoint MJ, Quintiere JG (2001) Predicting the piloted ignition of wood in the cone calorimeter using an integral model—effect of species, grain orientation and heat flux. Fire Saf J 36(4):391–415
22. Harada T (2001) Time to ignition, heat release rate and fire endurance time of wood in cone calorimeter test. Fire Mater 25(4):161–167
23. Shi L, Chew MYL (2013) Experimental study of woods under external heat flux by autoignition. J Therm Anal Calorim 111(2):1399–1407
24. Babrauskas V (2014) Ignition handbook—principles and applications to fire safety engineering, fire investigation, risk management and forensic science. Fire Science Publishers, Issaquah, WA
25. An W, Jiang L, Sun J, Liew KM (2015) Correlation analysis of sample thickness, heat flux, and cone calorimetry test data of polystyrene foam. J Therm Anal Calorim 119(1):229–238
26. Almirall JR, Furton KG (eds) (2004) Analysis and interpretation of fire scene evidence. CRC Press
27. Urakami H, Fukuyama M (1981) The influence of specific gravity on thermal conductivity of wood. Bulletin of the Kyoto Prefect Univ For (Japan) 25:38–45

Chapter 3
Methods of Calculation of Ignition Parameters

3.1 The Most Frequently Used Apparatus for the Measurement of Time to Ignition

There are several different methods that may be used for the measurement of time to ignition of flammable material samples dependent on the external heat flux. They are mostly composed of the following components:

- Sample holder
- Heat source
- Ignitor
- Combustion gas exhaust
- Components that measure other fire characteristics.

3.1.1 Ignition Cabinet

In their dissertation theses, Koohyar [1], Wesson [2], and Hallman [3] describe an apparatus for the observation of the impact of a flame heat flux on time to ignition of a sample. This apparatus was a chamber (Fig. 3.1) with two burners placed inside. A panel with a sample was placed between them. The sample was in a holder, which was placed inside the panel. Before the start of the test measurements, the sample was protected from each side by a radiation shield. A diffusion flame was produced in the burners using a liquid fuel. Combustion gases were transferred out of the chamber using an exhaust hood with a fan. The fan also ensured the flow of air into the chamber. On the lower side of the chamber, there was a metal honeycombed section and two fine mesh screens. One mesh was placed above the honeycomb and the other below it. This system was intended to reduce air turbulences. To prevent disturbance to the flame by the proximity of the chamber wall, guide panels were placed on the outer sides of the flames. The course of the test may have been observed through observation windows made of heat-resistant glass. The tests could be carried out with

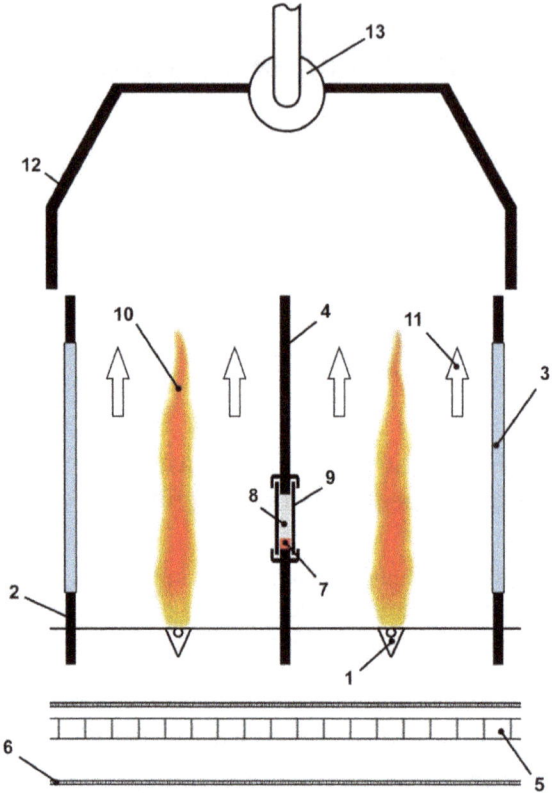

Fig. 3.1 Schematic representation of an ignition cabinet: 1—burner, 2—guide panel, 3—observation window, 4—sample panel, 5—honeycomb, 6—screen, 7—sample holder, 8—sample, 9—sample shield, 10—flame, 11—direction of air flow, 12—exhaust hood, 13—exhaust fan (based on Koohyar [1], Wesson [2], and Hallman [3])

only one side of the sample exposed to the flame or both. Wesson [2] and Hallman [3] also asserted that the flame could be replaced by high-temperature tungsten filament lamps.

3.1.2 Lateral Flame Spread Apparatus

This type of apparatus (Fig. 3.2) was described by Quintiere [4], Quintiere and Harkleroad [51], or Quintiere et al. [5]. It is also used in several standards, such as ISO 5658-2 Reaction to Fire Tests—Spread of Flame—Part 2: Lateral Spread on Building and Transport Products in Vertical Configuration (2020), or ASTM E1321 Standard Test Method for Determining Material Ignition and Flame Spread Properties (2018). Most often ignition is observed using horizontally placed samples. However,

3.1 The Most Frequently Used Apparatus for the Measurement … 71

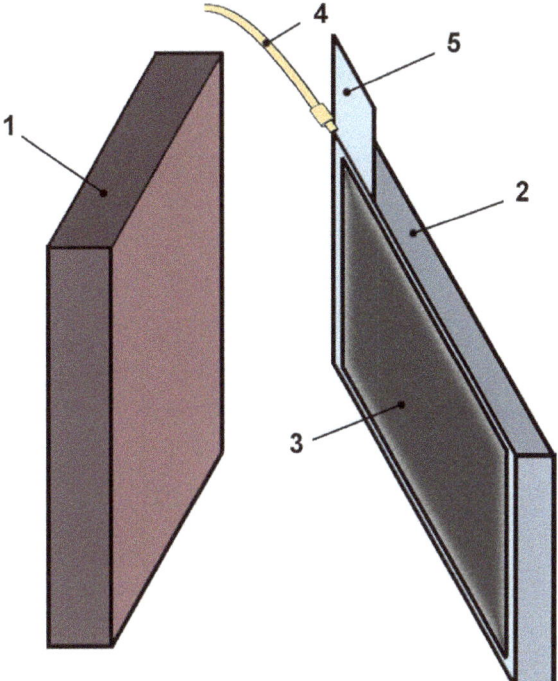

Fig. 3.2 Schematic representation of a lateral flame spread apparatus: 1—radiation panel, 2—sample holder, 3—sample, 4—burner, 5—steel plate (based on Quintiere and Harkleroad [51])

within this apparatus, the sample is positioned vertically. It has an elongated shape, with the length significantly exceeding the width. It is held by a frame, which is not parallel to the heat source. Thus, the external heat flux is not constant over the whole surface of the sample, but is applied according to a pre-defined function. This type of positioning is used for the measurement of the propagation of a flame. Flame ignition is provided by a flame situated at the edge of the sample. Quintiere and Harkleroad [51] state that if the ignitor were placed above the sample, it would be necessary to extend the plane of the sample surface upwards using a steel plate. This would enable the boundary layer containing the pyrolysed gases and the induced air flow to be maintained above the sample.

3.1.3 Fire Propagation Apparatus

A fire propagation apparatus (Fig. 3.3) is a type of laboratory calorimeter. It is an apparatus used to observe thermal phenomena. This apparatus is described in ASTM E2058 and ISO 12136 standards. It is made up of the following components [6, 7]:

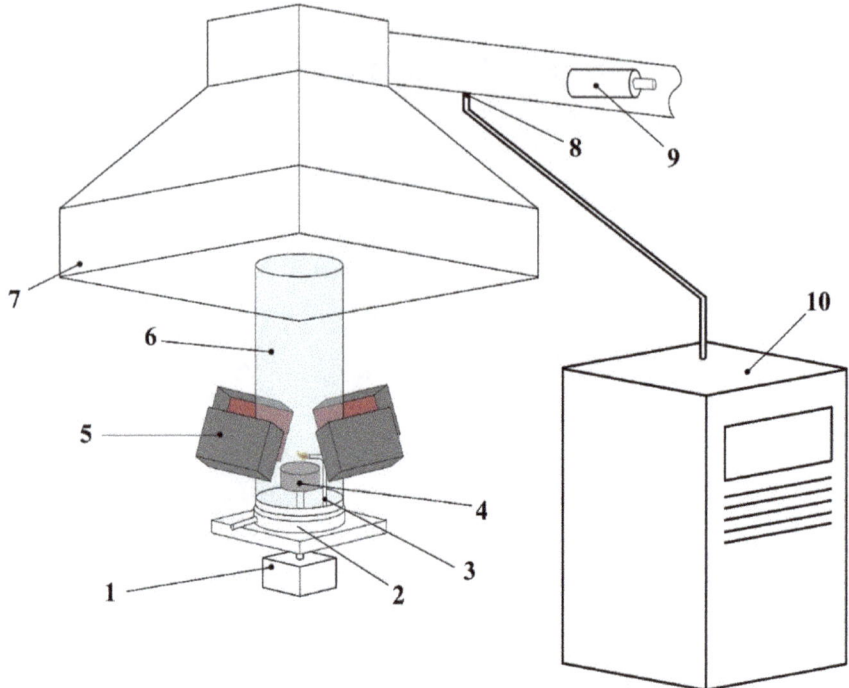

Fig. 3.3 Schematic representation of a fire propagation apparatus: 1—load cell system, 2—combustion air distribution system, 3—flame ignitor, 4—sample, 5—infrared panels, 6—quartz tube, 7—exhaust hood, 8—probes for collecting gaseous combustion products, 9—laser smoke measuring system, 10—combustion gas analyser

- An infrared heating system
- A load cell system
- An ignition pilot flame and timer
- A product gas analysis system
- A laser smoke measuring system
- A combustion air distribution system
- A water-cooled shield
- An exhaust system
- Test section instruments
- Calibration instruments
- A digital data acquisition system.

If the apparatus is used for the observation of ignition, the sample is placed horizontally and subsequently exposed to a stable external heat flux. Four infrared heat sources with tubular tungsten filament quartz lamps are used as heat sources. In order to achieve maximum absorptivity, the surface of the sample is coated with a thin layer of black paint. The ignition of the products produced by the thermal degradation

3.1 The Most Frequently Used Apparatus for the Measurement … 73

of the sample is provided by a flaming ignitor, which ignites the air/ethylene mixture. During the test measurements, the sample is within a quartz pipe, which allows the flow of the atmosphere around the sample to be controlled using a combustion air distribution system. The weight loss during the test is observed using a scale with a measurement range of 0–1000 g, with an accuracy of 0.1 g. In addition to the time to ignition, it is equally possible to observe the rate of heat release, the composition of the combustion products, and the optical density of the resultant smoke. The Fire Propagation Test Method involves the use of vertical specimens that ignite near the base of the specimen after application of an external radiant heat flux and pilot flame. A shield composed of aluminium cylinders with water circulating inside them was used to prevent the exposure of the sample to the heat flux until the heat sources had stabilised [6, 7].

3.1.4 Ignitability Test Apparatus

As the name suggests, an ignitability apparatus is specifically intended to measure the time to ignition of samples. It is described in: ISO 5657:1997 [8] Reaction to Fire Tests—Ignitability of Building Products Using a Radiation Heat Source. The heat source is a cone heater capable of generating an even, constant heat flux ranging from 10 to 70 kW m^{-2} on a specific area on the surface of the sample. The sample itself is placed horizontally on a pressing plate. It should be a square of 165 mm × 165 mm with a maximum thickness of 70 mm. A baseboard made of non-flammable insulating material with a thickness of 6 mm is placed underneath the sample. Together with the sample, it is wrapped in aluminium foil with a circular central cut-out of 140 mm diameter. Hence, an area of only 154 cm^2 is exposed to the external heat flux rather than the whole upper surface of the sample. The force necessary to push the sample against the masking plate is applied by a mechanism with an adjustable counterweight. A specimen screen plate, placed over the top of the masking plate, ensures the sample is protected from the heat flux until the start of the test measurements. The ignition source for the gaseous combustion products released from the sample is provided by a gas burner. The apparatus also includes a mechanism that moves the burner from an inactive position to the normal test position at regular intervals (4 s). A schematic representation of an ignitability test apparatus is shown in Fig. 3.4.

3.1.5 Cone Calorimeter

A cone calorimeter (Fig. 3.5) is especially used for the measurement of the rate of heat release. It is described in several standards, including ASTM E1354 and ISO 5660. A schematic representation of a cone calorimeter is shown in Fig. 3.5.

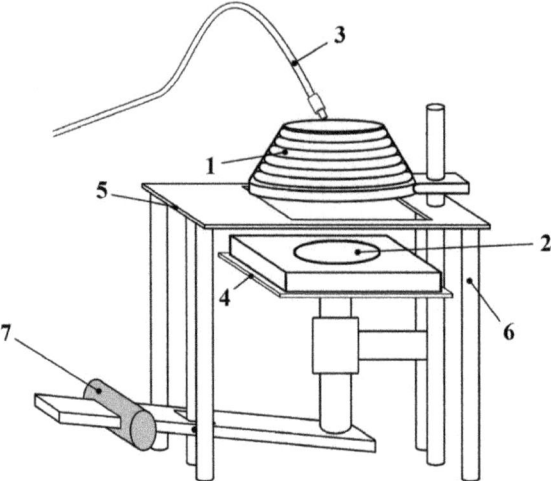

Fig. 3.4 Schematic representation of an ignitability test apparatus: 1—cone heater, 2—sample, 3—burner, 4—pressing plate, 5—masking plate, 6—support framework, 7—counterweight

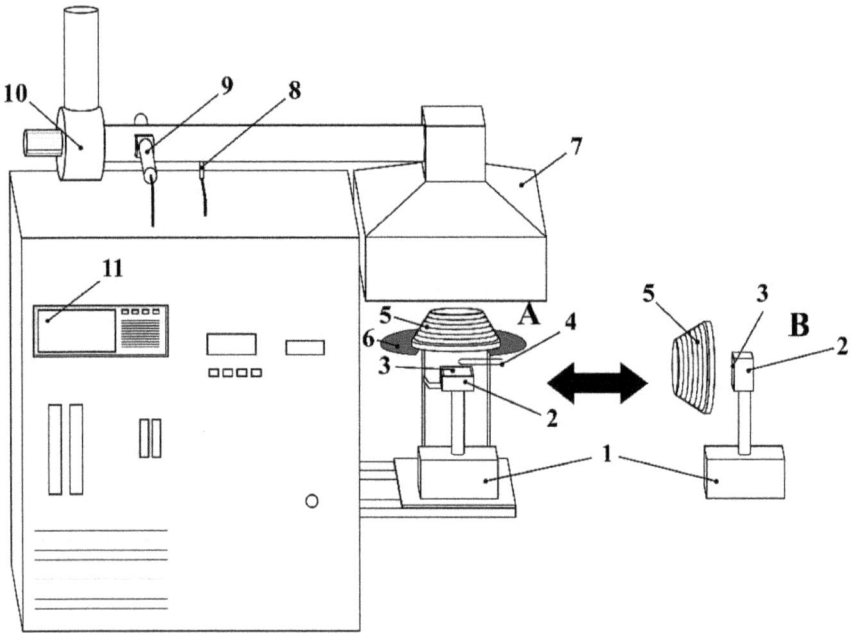

Fig. 3.5 Cone calorimeter: 1—scales, 2—sample holder, 3—sample, 4—spark ignitor, 5—cone heater, 6—radiation shield, 7—exhaust hood, 8—probe for the collection of samples of combustion gases, 9—optometric system for measuring smoke density, 10—fan, 11—gas analyser. **a** Horizontal orientation, **b** vertical orientation

The apparatus was named after one of its fundamental components, a cone heater. The heat source is designed in the shape of a truncated cone, and this is intended to ensure that the entire surface of the sample is exposed to a practically constant density of external heat flux. Owing to its truncated shape, the products of thermal degradation and combustion may escape directly upwards during measurements. A movable radiation shield is situated below the heat source. It is intended to retain the external heat flux at the time that the sample is inserted into the calorimeter. The sample itself measures 100 mm × 100 mm and is 50 mm thick. Before any measurements, all the sides that should not be exposed to the external heat flux are wrapped in aluminium foil, and subsequently, it is placed into the holder to secure it on the scale during measurement. A thermally insulating layer of mineral wool is placed underneath the sample, in a holder. An electrical spark ignitor is provided to ignite the gaseous products released by the sample. During the tests, it is situated between the surface of the sample and the cone heater. The products of thermal degradation and combustion gases are captured by an exhaust hood and subsequently extracted through the extraction pipe by a fan. There is also a probe for the collection of samples of the combustion gases in this extraction pipe. It is in the shape of a perforated circular metal pipe. The collected samples are transported to the combustion gas analyser. Apart from the probe that collects combustion gases, an optometric system for measuring the optical density of smoke may also be placed on the extraction pipe, after the probe.

The heater included in a cone calorimeter is usually designed to be able to apply a heat flux of up to 100 kW m^{-2} on the surface of the sample. However, lower heat fluxes tend to be used for testing, principally 50 kW m^{-2}.

A sample may be placed horizontally or vertically (Fig. 3.5a, b). The difference between these modes of testing, apart from the orientation of the sample, lies in the holder used, the orientation of the heater, and the placement of the spark ignitor. The results obtained through measurements taken in the horizontal or vertical position may differ substantially.

A comparison of the cone calorimeter and ignitability test apparatus was conducted by Östman and Tsantaridis [9]. Based on the comparison of the time to ignition data from each apparatus, they state that the data seems to agree fairly well or at least rank the different materials in approximately the same order. Thus, in most cases, only one of the test procedures is necessary.

3.2 The Calculation of Ignition Parameters

3.2.1 Critical Heat Flux

The critical density of heat flux (often shortly referred to as critical heat flux) means the lowest heat flux density that reaches the surface of a material which is able to trigger the ignition of permanent flaming combustion in a sample. This parameter may be determined for methods with a combustion ignitor (such as a spark or a flame) as well as without an ignitor. Critical heat flux without an ignitor may also be referred to as the critical autoignition heat flux and returns higher values than measurements with an ignitor (critical flash heat flux).

As the formulae in the section "The Correlation between External Heat Flux and Time to Ignition" clearly show, the time to ignition of an object is a function of the heat flux that reaches its surface. In general, this correlation may be expressed as follows:

$$\left(\frac{1}{t_{ig}}\right)^{\frac{1}{n}} = A_q q_e + B_q \qquad (3.1)$$

A_q and B_q are constants corresponding to the ignition of a material under test conditions and the coefficient n has a positive value, especially in the interval of 1 to 2. Provided that the critical heat flux represents a border between the ignition and non-ignition of a sample, it may be referred to as the value of the external heat flux under which a sample will ignite in an infinite period of time. Hence, it may be calculated as the limit of:

$$q_{cr} = \lim_{t \to \infty} \frac{1}{A_q t_{ig}^{\frac{1}{n}}} - \frac{B_q}{A_q} = -\frac{B_q}{A_q} \qquad (3.2)$$

Based on the values measured, it is possible to create a correlation chart of external heat flux vs the inverse values of the corresponding time to ignition to the power of $\frac{1}{n}$. Consequently, the critical heat flux may be obtained at the overlap of the extrapolation of the trend line with the axis that corresponds to the applied external heat flux (Fig. 3.6).

Spearpoint and Quintiere [10] explain that the critical heat flux determined in this way is the intercept heat flux, and the actual value of the critical heat flux may differ. For wood, they propose a coefficient of 0.76:

$$q_{cr} = \frac{q_{intercept}}{0.76} \qquad (3.3)$$

Selected correlations between heat flux and time to ignition are summarised in Table 3.1. As mentioned before, the most commonly used n-values range from 1 to 2. Mikkola and Wichman [11] suggest that these two values are the limits for thermally thin and thermally thick materials. However, some authors propose even higher values.

Tsai [25] states that for samples tested vertically, the critical heat flux is 15% higher than for horizontally oriented samples:

$$q_{cr(V)} = 1.15 q_{cr(H)} \qquad (3.4)$$

3.2 The Calculation of Ignition Parameters

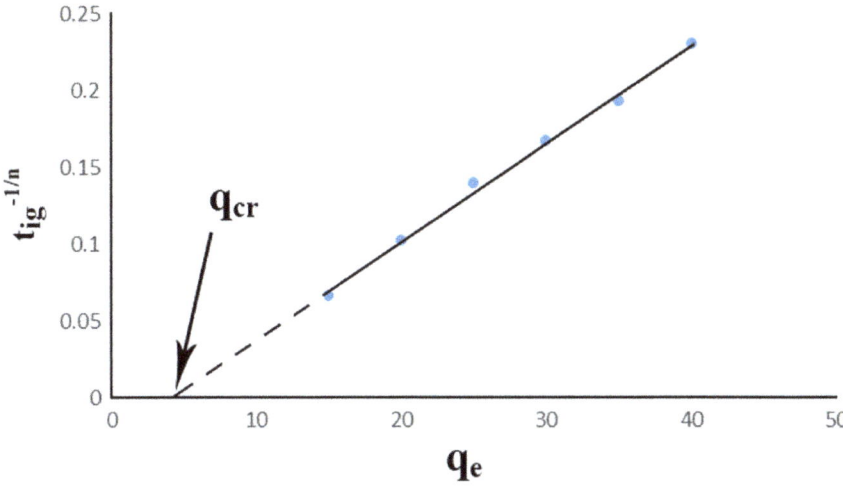

Fig. 3.6 The determination of critical heat flux

Table 3.1 Examples of the n coefficient as proposed by various literary sources

n coefficient	Use	Source
1	Wood	[2]
	Plasterboard, flash	[12]
	Wood, flash	
	Thermally thin materials	[11]
1.25	Wood, autoignition	[13]
1.5	Wood, flash	[13]
	Thermally medium materials	[11]
1.82	Wood	[14]
	Wood	[15]
1.83	Wood, flash	[16]
2	Rubber and plastics	[3]
	Solid flammable materials	[4]
	Polypropylene and polymethylmethacrylate	[17]
	Thermally thick materials	[11]
	Polymethylmethacrylate	[18]
	Wood	[10]
	Foam polystyrenes	[19]
2.8	Wood	[20]
3	Wood	[21]
n	Mathematical model	[22]
	Cellulosic materials	[23]
	General formula	[24]

$q_{cr(V)}$ is the critical heat flux in the vertical position, and $q_{cr(H)}$ is the critical heat flux in the horizontal position.

3.2.2 Flux–Time Product

As already mentioned in Sect. 3.2, "The Correlation between External Heat Flux and Time to Ignition", one of the parameters used to express the ignition of flammable materials is the flux–time product. Baker et al. [26] state that the theory behind the FTP concept is that when a material is subjected to an external flux, the FTP aggregates until it exceeds a threshold, and the material ignites. Essentially, the FTP is a quasi-material constant which represents a specific combination of thermal properties that also accounts for the thermal thickness of the sample. It can be determined through the application of Formula (2.30) above. Baker et al. [26] propose this version:

$$q_e = \frac{(\text{FTP})^{\frac{1}{n}}}{t_{ig}^{\frac{1}{n}}} + q_{cr} \tag{3.5}$$

Although the concept of FTP was proposed for the determination of time to ignition in thermally closed systems, it may also be useful for open systems [27]. Shields et al. [28] state that the index of the flux–time product of open systems is greater, up to 1.

3.2.3 The Thermal Response Parameter

Tewarson and Odgen [29] describe the correlation between the time to ignition and external heat flux using the thermal response parameter (TRP) which represents the combination of the thermal conductivity, density, specific heat, and the difference between the ignition temperature and ambient temperature. They expressed it mathematically as:

$$\text{TRP} = \sqrt{K\rho c}(T_p - T_0) \tag{3.6}$$

The thermal response parameter is an indicator of the ignition resistance of a material [30]. A high TRP means that heating up, ignition, and the initiation of fire take longer [31].

Provided that the material is thermally thick, the following formula may be used to calculate the TRP [29]:

3.2 The Calculation of Ignition Parameters

$$\frac{1}{\sqrt{t_{ig}}} = \sqrt{\frac{4}{\pi} \frac{q}{\text{TRP}}} \quad (3.7)$$

TRP may then simply be calculated based on the value of the slope of the trend line:

$$\text{TRP} = \sqrt{\frac{4}{\pi} \frac{1}{\text{Slope}}} \quad (3.8)$$

3.2.4 The Transfer Convective Coefficient

The transfer convective coefficient for horizontally placed samples may be calculated using Janssens' method as follows [32, 33] (Dao [31]):
For heat fluxes lower than 50 kW m^{-2}:

$$h_c = 0.01198 + 3.74 \times 10^{-4} q_e \quad (3.9)$$

For heat fluxes of 50 kW m^{-2} and greater:

$$h_c = 0.0255 + 6.5 \times 10^{-5} q_e \quad (3.10)$$

For heat fluxes from 20 to 65 kW m^{-2} for cone calorimeter measurements, Dietenberger proposes the following correlation [34]:

$$h_c = 0.01433 + 1.33 \times 10^{-4} q_e \quad (3.11)$$

3.2.5 The Apparent Thermal Inertia and Thermal Diffusivity

Thermal inertia expresses how fast the temperature increases at the surface of a material when it is exposed to heat. Materials with lower thermal inertia start to burn more quickly than those with higher thermal inertia [35]. It is often defined by the mathematical formula:

$$I = K\rho c \quad (3.12)$$

I is the thermal inertia. Since it forms a part of the parameter of a thermal response, it may be calculated based on the following formula:

$$I = \left(\frac{TRP}{T_p - T_0}\right)^2 \tag{3.13}$$

Thermal inertia depends on temperature, since its value at the moment of ignition is not identical to its value under normal circumstances. Its apparent value can be determined based on ignition data [10].

If we know the thermal inertia at a certain temperature (e.g. at ambient temperature), it is possible to estimate its approximate value at a higher temperature [35].

$$I_{T_2} \approx K_{T_1} \rho_{T_1} c_{T_1} \frac{T_2}{T_1} \tag{3.14}$$

I_{T_2} is the thermal inertia at T_2, K_{T_1} is the heat conductivity at T_1, ρ_{T_1} is the density at T_1, and c_{T_1} is the thermal capacity at T_1.

Thermal diffusivity describes the time-dependent, non-steady-state aspects of heat flow [36]. It may be calculated using the same characteristics required for the calculation of thermal inertia:

$$\kappa = \frac{K}{\rho c} \tag{3.15}$$

κ is thermal diffusivity. Similar to thermal inertia, thermal diffusivity may also be determined for various temperatures [35]:

$$\kappa_{T_2} = \frac{K_{T_1} T_2}{\rho_{T_1} c_{T_1} T_1} \tag{3.16}$$

This parameter is important from the perspective of the thermal penetration depth (δ). Thermally thin materials have a lower thermal thickness, and thermally thick materials have a higher thermal thickness than the penetration depth. Its value may be calculated as follows:

$$\delta = A_\delta \sqrt{\alpha\, t_{ig}} \tag{3.17}$$

A_δ is a coefficient that has been given various values by individual authors. In the Handbook of Building Materials for Fire Protection, its value is 1 [35]. But even values such as 4, 1.13, 1.2, 2, or 2.45 may be used [15, 37, 38]. Values of thermal inertia and thermal diffusivity for selected polymers are shown in Table 3.2.

In some cases, the Biot number is used to determine the thermal thickness of the material:

$$B_i = \frac{h_c . L_0}{K} < 0.1 \tag{3.18}$$

3.2 The Calculation of Ignition Parameters

Table 3.2 Thermal inertia and thermal diffusivity of selected polymers

Polymer	I at 298 K [kW^2 s m^{-4} K^{-2}]	κ [m^2 s^{-1} × 10^7]	Source
ABS	0.41	1.65	[35]
HIPS	0.31	1.54	[35]
PBT	0.48	1.01	[35]
PC	0.29	1.36	[35]
PE HD	0.82	2.24	[35]
PE LD	0.54	2.65	[35]
PE MD	0.63	2.53	[35]
PET	0.59	1.29	[35]
PETG	0.59	1.42	[35]
PMMA	0.33	1.19	[35]
PP	0.25	0.89	[35]
PS	0.18	1.04	[35]
PTFE	0.56	1.11	[35]
PU	0.44	0.99	[35]
PUR	0.37	0.98	[35]
PVC (flex)	0.29	0.98	[35]
PVC (rigid)	0.26	1.34	[35]
PETG	0.45	1.87	[39]
Wood (dry)			
Redwood		1.85	[34]
Northern white cedar		2.0	[40]
Red pine		1.9	[40]
Loblolly pine		1.8	[40]
Yellow birch		1.8	[40]
White oak		1.8	[40]

B_i is the Biot number, and L_0 is the thickness of the material. Under these circumstances, the temperature of the material will increase uniformly [41].

Thermal thickness may equally be specified based on correlations typical of certain groups of materials. For example, Babrauskas and Parker [42] suggest the following for particle board:

$$\delta = 0.6 \frac{\rho}{q_e} \qquad (3.19)$$

the ratio of density to the heat flux reaching the surface is derived from general principles of heat conductivity, and the 0.6 constant was obtained through measurement.

Chen et al. similarly determined a correlation for flame-retardant ethylene-propylene-diene monomer rubber [43]:

$$\delta = 0.0772 \frac{\rho}{q_e} \quad (3.20)$$

Shi and Chew [14] suggest the following version of this formula for polymers under autoignition conditions:

$$\delta = 0.14 \frac{\rho}{q_e} \quad (3.21)$$

3.2.6 The Ignition Temperature

Ignition temperature is relatively frequently quoted as a characteristic of materials. One of the possible reasons for this is that it is quite easily understood, even by laymen. The different methods used to calculate it are generally based on time to ignition under external heat flux, and they especially differ according to the assumed simplifications. It is possible to estimate an approximate value based on critical heat flux according to the Stefan–Boltzmann law. If the surface of the material is a black body, then:

$$T_{bb} = \left(\frac{q_{cr}}{\sigma} \right)^{\frac{1}{4}} \quad (3.22)$$

T_{bb} is the temperature that corresponds to a black body. However, in realistic conditions, the surface of a material does not behave as a black body. Still, it may have similar properties under specific conditions. For this reason, the authors recommend that the surface of the sample be painted using matte black paint that applies a thin layer of carbon [18, 44]. Simms [45, 46] describes the impact of colouring the surface of the sample on the ignition of lignocellulosic materials—the darker it is, the shorter the time to ignition. The absorptivity of the surface of the material at a low heating rate is independent of heating rate. The surface that is exposed to the external heat flux gets progressively darker; thus, absorptivity is a function of time. For spontaneous ignition of most cellulosic materials, the effective absorption rate for a sample that is exposed to a radiation panel is near unity. The effective absorption may differ by up to 60% depending on the time for which the heat flux is applied. As the time to ignition increases, the effective rate of absorption gradually becomes a constant [47]. The value of effective absorption may also be expressed depending on the irradiation that reaches the surface (Table 3.3).

The effective absorptivity of products exposed to radiation from fires is typically in the interval 0.75–0.95 [48].

3.2 The Calculation of Ignition Parameters

Table 3.3 Effective absorptivity of various materials in a cone calorimeter [48]

q [kW m^{-2}]	10	25	50	75	100
Material	Effective absorptivity [%]				
Wood products					
Plywood	86	84	81	79	76
Light lacquered ash tree flooring	90	88	86	84	82
Light non-lacquered ash tree flooring	86	84	81	80	77
Medium dark lacquered oak flooring	91	89	87	85	83
Medium dark non-lacquered oak flooring	86	84	82	80	77
Plastics					
ABS (white)	91	90	88	86	84
ABS (black)	92	92	92	92	92
PE (nature)	93	93	93	93	93
PE (yellow)	93	92	92	91	90
PE (black)	93	93	93	93	93
PP (grey)	92	92	91	91	90
PTFE (nature)	84	78	73	70	66
PVC (clear)	91	90	88	86	84
PVC (white)	91	89	87	85	82
PVC (white, foamed)	82	82	78	76	73
PVC (grey)	91	90	90	90	90
PVC (black)	93	93	93	93	93

If the absorptivity value of a material is known, it is possible to use the Stefan–Boltzmann law for a grey body. The time to ignition may then be calculated as:

$$T_{ig} = \left(\frac{\alpha_{abs} q_{cr}}{\sigma}\right)^{\frac{1}{4}} \qquad (3.23)$$

α_{abs} is absorptivity. Formula (3.23) only applies if the surface of the material is heated by radiation heat flux, all the heat applied is absorbed by the surface of the sample and exclusively used to heat up the thin upper layer and it is not reflected into the environment. Conduction heat into the sample is also ignored, and the surface material is considered to be inert.

Taking convective heat into consideration, ignition temperature may also be determined based on the often-proposed formula [37]:

$$q_e = \frac{1}{\varepsilon}\left[h_c(T_{ig} - T_0) + \varepsilon \sigma T_{ig}^4\right] \equiv q_{cr} \qquad (3.24)$$

Delichatsios proposes the use of this formula without emissivity (a black body assumption), with the value of h_c for horizontally oriented surfaces being approximately 5 W m^{-2} K^{-1} (Delichatsios [52]).

For the calculation of critical heat flux, the sum of heating losses includes not only convection and radiation from the surface of the sample, but also losses from the rear face and the endothermicity of the decomposition process. Then, for piloted ignition of an infinite slab [49]:

$$\alpha_a q_{cr} = h_c (T_{ig} - T_0) + \varepsilon \sigma T_{ig}^4 + L_v m_{cr} + K \frac{T_{ig} - T_r}{L_0} \qquad (3.25)$$

α_{aba} is the absorptity, L_v is the heat of gasification, m_{cr} is the mass flux of fuel vapours, and T_r is the temperature of the rear face.

Estimating the ignition temperature of various materials is also possible using other methods. For wood, Buschman suggests using a correlation between the ignition temperature and thermal inertia [50]:

$$T_{ig} = 667 - 0.527 K \rho c \times 10^9 \qquad (3.26)$$

It must be noted that this correlation was derived for cal, cm, g, and °C.

References

1. Koohyar AN (1968) Ignition of wood by flame radiation. The University of Oklahoma, Oklahoma
2. Wesson HR (1970) The piloted ignition of wood by radiant heat. The University of Oklahoma, Oklahoma
3. Hallman JR (1971) Ignition characteristics of plastics and rubber. The University of Oklahoma, Oklahoma
4. Quintiere J (1981) A simplified theory for generalizing results from a radiant panel rate of flame spread apparatus. Fire Mater 5(2):52–60
5. Quintiere J, Harkleroad M, Walton D (1983) Measurement of material flame spread properties. Combust Sci Technol 32(1–4):67–89
6. ASTM E2058-19:2019 (2019) Standard test methods for measurement of material flammability using a fire propagation apparatus (FPA)
7. ISO 12136:2011 (2011) Reaction to fire tests—measurement of material properties using a fire propagation apparatus
8. ISO 5657:1997. Reaction to fire tests—ignitability of building products using a radiant heat source
9. Östman B, Tsantaridis L (1990) Ignitability in the cone calorimeter and the ISO ignitability test
10. Spearpoint MJ, Quintiere JG (2001) Predicting the piloted ignition of wood in the cone calorimeter using an integral model—effect of species, grain orientation and heat flux. Fire Saf J 36(4):391–415
11. Mikkola E, Wichman IS (1989) On the thermal ignition of combustible materials. Fire Mater 14(3):87–96
12. Bluhme DA (1987) ISO ignitability test and proposed criteria. Fire Mater 11(4):195–199

References

13. Lawson DI, Simms UD (1952) The ignition of wood by radiation. Br J Appl Phys 3(9):288
14. Shi L, Chew MYL (2013) Fire behaviors of polymers under autoignition conditions in a cone calorimeter. Fire Saf J 61:243–253
15. Babrauskas V (2014) Ignition handbook—principles and applications to fire safety engineering, fire investigation, risk management and forensic science. Fire Science Publishers, Issaquah, WA
16. Janssens M (1991) Fundamental thermophysical characteristics of wood and their role in enclosure fire growth. Doctoral dissertation, Ghent University
17. Hu X, Clark FR (1988) The use of the ISO/TC92 test for ignitability assessment. Fire Mater 12(1):1–5
18. Delichatsios MA, Panagiotou TH, Kiley F (1991) The use of time to ignition data for characterizing the thermal inertia and the minimum (critical) heat flux for ignition or pyrolysis. Combust Flame 84(3–4):323–332
19. An W, Jiang L, Sun J, Liew KM (2015) Correlation analysis of sample thickness, heat flux, and cone calorimetry test data of polystyrene foam. J Therm Anal Calorim 119(1):229–238
20. Wesson HR, Welker JR, Sliepcevich CM (1971) The piloted ignition of wood by thermal radiation. Combust Flame 16(3):303–310
21. Harada T (2001) Time to ignition, heat release rate and fire endurance time of wood in cone calorimeter test. Fire Mater 25(4):161–167
22. Smith EE, Satija S (1983) Release rate model for developing fires
23. Toal BR, Silcock GWH, Shields TJ (1989) An examination of piloted ignition characteristics of cellulosic materials using the ISO ignitability test. Fire Mater 14(3):97–106
24. Zhang J, Shields TJ, Silcock GWH (1996) Fire hazard assessment of polypropylene wall linings subjected to small ignition sources. J Fire Sci 14(1):67–84
25. Tsai KC (2009) Orientation effect on cone calorimeter test results to assess fire hazard of materials. J Hazard Mater 172(2–3):763–772
26. Baker G, Collier P, Wade C, Spearpoint M, Fleischmann CM, Frank K, Sazegara S (2013) A comparison of a priori modelling predictions with experimental results to validate a design fire generator submodel. Proc Fire Mater 449–460
27. Silcock GWH, Shields TJ (1995) A protocol for analysis of time-to-ignition data from bench scale tests. Fire Saf J 24(1):75–95
28. Shields TJ, Silcock GW, Murray JJ (1994) Evaluating ignition data using the flux time product. Fire Mater 18(4):243–254
29. Tewarson A, Ogden SD (1992) Fire behavior of polymethylmethacrylate. Combust Flame 89(3–4):237–259
30. Xu Q, Chen L, Harries KA, Zhang F, Liu Q, Feng J (2015) Combustion and charring properties of five common constructional wood species from cone calorimeter tests. Constr Build Mater 96:416–427
31. Dao DQ, Luche J, Richard F, Rogaume T, Bourhy-Weber C, Ruban S (2013) Determination of characteristic parameters for the thermal decomposition of epoxy resin/carbon fibre composites in cone calorimeter. Int J Hydrogen Energy 38(19):8167–8178
32. Janssens ML (1993) Improved method of analysis for the LIFT apparatus. Part I: inflammation, 2nd fire and material conference. Interscience Communication, London, 23–24 Sept 1993, pp 37–46
33. Batiot B, Luche J, Rogaume T (2014) Thermal and chemical analysis of flammability and combustibility of fir wood in cone calorimeter coupled to FTIR apparatus. Fire Mater 38(3):418–431
34. Dietenberger MA (1996) Ignitability analysis using the cone calorimeter and LIFT apparatus. In: Proceedings of the international conference on fire safety, July 22–26, 1996, Columbus, OH, vol 22. Product Safety Corporation, pp 189–197
35. Harper CA (2004) Handbook of building materials for fire protection. McGraw-Hill, New York, pp 384–385
36. Bicerano J (2002) Prediction of polymer properties, 3rd edn. CRC Press, New York

37. Rhodes BT, Quintiere JG (1996) Burning rate and flame heat flux for PMMA in a cone calorimeter. Fire Saf J 26(3):221–240
38. Lautenberger CHRIS, Fernandez-Pello AC (2005) Approximate analytical solutions for the transient mass loss rate and piloted ignition time of a radiatively heated solid in the high heat flux limit. Fire Saf Sci 8:445–456
39. Sd3D (a), PETG (polyethylene terephthalate copolymer). Technical data sheet
40. TenWolde A, McNatt JD, Krahn L (1988) Thermal properties of wood and wood panel products for use in buildings (No. DOE/OR/21697-1). Forest Service, Madison, WI (USA). Forest Products Lab
41. Drysdale DD (1986) Fundamentals of the fire behaviour of cellular polymers. In: Fire and cellular polymers. Springer, Dordrecht, pp 61–75
42. Babrauskas V, Parker WJ (1987) Ignitability measurements with the cone calorimeter. Fire Mater 11(1):31–43
43. Chen R, Lu S, Li C, Ding Y, Zhang B, Lo S (2016) Correlation analysis of heat flux and cone calorimeter test data of commercial flame-retardant ethylene-propylene-diene monomer (EPDM) rubber. J Therm Anal Calorim 123(1):545–556
44. Tewarson A (2002) Generation of heat and chemical compounds in fires. In: SFPE handbook of fire protection engineering, pp 82–161
45. Simms DL (1960) Ignition of cellulosic materials by radiation. Combust Flame 4:293–300
46. Simms DL (1964) On the spontaneous ignition of cellulosic materials by radiation. Fire research note no 573
47. Simms DL, Law M, Hinkley PL (1957) The effect of absorptivity on the ignition of materials by radiation. Fire Saf Sci 308:1–1
48. Försth M, Roos A (2011) Absorptivity and its dependence on heat source temperature and degree of thermal breakdown. Fire Mater 35(5):285–301
49. Thomson HE, Drysdale DD, Beyler CL (1988) An experimental evaluation of critical surface temperature as a criterion for piloted ignition of solid fuels. Fire Saf J 13(2–3):185–196
50. Buschman AJ (1961) Ignition of some woods exposed to low level thermal radiation. National Bureau of Standards, Washington DC
51. Quintiere JG, Harkleroad M (1985) New concepts for measuring flame spread properties, MBSIR 84-2943. U.S. Department of commerce, National bureau of Standards, National Engineering Laboratory, Center for Fire Research, Gaithersburg, MD 20899
52. Delichatsios MA (2000) Ignition times for thermally thick and intermediate conditions in flat and cylindrical geometries. Fire Saf Science 6:233–244

Chapter 4
Comparing the Ignition Parameters of Various Polymers

4.1 Materials

Samples of flat material were used for the measurements. From the synthetic polymers, we selected plastics, and from the natural polymers, we chose wooden materials. The sample dimensions were 100 mm × 100 mm with a thickness that differed depending on the material.

4.1.1 Plastics

The plastic samples were selected based on the possibility to 3D print them. This method of production of various objects has become increasingly popular over the last few years. There is a wide range of different 3D printing methods. The most frequently used include [1]:

1. Fused deposition modelling (FDM)—a solid polymer filament is melted and subsequently applied in layers using a jet. Thermoplastic polymers with a melting point of 250–300 °C are used. This is the most popular method for 3D printing owing to the low cost of material and simple printer design.
2. Vat polymerisation including stereolithography (SLA) and direct light processing (SLS)—a photopolymer resin is exposed to a precisely controlled source of light, which causes it to solidify. Since the energy may be concentrated over very small areas, this method allows high print resolution. However, this method is both slower and more expensive than FDM.
3. Powder bed fusion—a powder-based polymer is melted and subsequently solidified. The most common use of this method is in selective laser sintering (SLS).

The test samples were printed using FDM. The jet orifice was 0.4 mm in diameter, and a single layer had a thickness of 0.35 mm. The samples were 8 mm thick; the

Table 4.1 Samples of synthetic polymers and processing temperatures

Reference	Material	Colour	ρ [g cm^{-3}]	Nozzle temperature [°C]	Bed temperature [°C]
F1	ABS-T	Transparent	1.08	255	110
F2	ABS-T	Black (RAL9011)	1.08	255	110
F3	PLA	Transparent	1.24	215	60
F4	PLA	Fluorescent yellow	1.24	215	60
F5	PETG	Transparent brown (RAL 8015)	1.27	250	90
F6	PETG + phoslite	Black (RAL 9011)	1.27	250	90

fill was set to a density of 20% with a cubic pattern. Three polymer materials were selected, and two types of filaments were used from each to print the surfaces of the samples:

1. Polylactic acid (PLA)
   (a) Transparent filament
   (b) Yellow filament
2. Acrylonitrile butadiene styrene copolymer with methylmethacrylate (ABS-T)
   (a) Transparent filament
   (b) Black filament
3. Poly(ethyleneterephthalate)-glycol (PETG)
   (a) Transparent brown filament
   (b) Black filament with fire retardant.

For the PLS and ABS-T samples, the surface was made up of four layers and for PETG, three layers. The characteristics of the samples and the processing temperatures are given in Table 4.1.

4.1.2 Wood-Based Materials

The wood-based materials chosen included samples of two species of solid wood, one of a thermally modified wood and two types of wood composites. The solid wood samples included a coniferous wood species (pine) as well as a hardwood (ash). The thermally modified wood was thermowood, and the wood composites

4.1 Materials

Table 4.2 Characteristics of the samples of wood-based materials

Material	ρ [kg m^{-3}]	L_0 [mm]	n_{por} [%]	Average colour coordinates		
				L^*	a^*	b^*
Pine wood	470	18	68.7	59.3	16.5	30.9
Ash wood	676	19	54.9	75.2	8.6	21.6
Thermowood	432	19	71.2	49.4	14.2	27.1
Plywood	647	18	56.9	73.4	12.0	33.1
Blockboard	567	17	62.2	72.9	9.7	20.1

were plywood and blockboard. The plywood was composed of boards of soft wood, with the outer layers being made of pine. The blockboard had a similar composition to the plywood—with a core of soft wood and a surface that comprised a combination of alder and birch. Prior to test measurements, all samples were oven dried for 3 days at a temperature of 100 °C and left to cool in a desiccator. The average density and colour of the samples of the wood-based materials are given in Table 4.2.

Since the cell walls of almost all wood types have the same density (around 1500 m^3), wood density is given by the ratio of the cell walls to the cell lumens [2]. Plötze and Niemz [3] suggest specific (cell wall) density in the range of 1451–1528 kg m^{-3} with an average value of 1493 kg m^{-3}, which corresponds very well with the indicated data. Based on this value, it is possible to calculate wood porosity (n_{por}) as follows:

$$n_{por} = \frac{1500 - \rho}{1500} \cdot 100\% \quad (4.1)$$

The mean porosity of individual materials is provided in Table 4.2. Pine has a higher porosity than ash. The calculated values may differ slightly for thermowood, plywood, and blockboard. In the case of thermowood, this is due to thermal degradation of the cell walls triggered by thermal changes which have a critical impact on their properties [4, 5]. The porosity of plywood and blockboard may be influenced by the quantity of glue used in their production.

Wood is a natural material formed by processes within a living organism, and during its growth, it is affected by the surrounding environment (e.g. mechanical stresses and weather conditions). Hence, the properties of different samples may show significant differences within a single type of wood or even within a single tree. Based on the comparisons made using box plots showing the bulk density of samples of wood-based materials (Fig. 4.1), it appears clear that the lowest covariance was seen in the pine samples (almost 7%) and the highest in blockboard (around 17.5%).

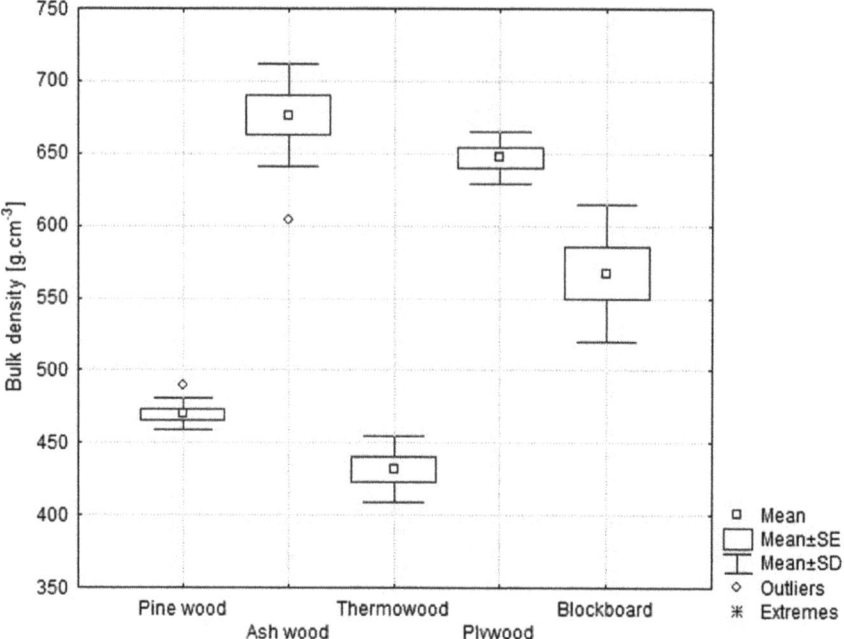

Fig. 4.1 Bulk density of measured samples

4.2 Methods Used to Compare the Ignition Parameters of Different Polymers

A cone calorimeter, as described in detail in the section "Methods of Calculation of Ignition Parameters", was used as the measurement apparatus. The samples were placed into the sample holder, the edges that were not to be exposed to the heat flux were wrapped in aluminium foil to protect them, and then they were placed under the cone heater. The measurements were conducted horizontally, and a spark ignitor was used. A visual observation of the time to ignition was made and manually recorded using a stopwatch. The gas flow rate in the exhaust pipe was set to 0.024 m^3 s^{-1}. Seven different heat fluxes, in 5 kW m^{-2} increments, from 25 to 55 kW m^{-2} were applied. The ambient temperature, air humidity (H_a), and atmospheric pressure (p_{atm}) during the measurements are given in Table 4.3.

Table 4.3 Ambient conditions during the tests

Material	T_0 [°C]	H_a [%]	p_{atm} [kPa]
F1	22.1	31–33	99.78–100.21
F2	22.1	31–33	99.77–100.21
F3	22.4	31–33	99.78–100.19
F4	22.3	31–33	99.77–100.21
F5	22.0	30–33	99.78–100.16
F6	22.4	30–33	99.79–100.23
Pine	22.1	31–33	100.23–100.27
Ash	22.3	31–33	100.10–100.23
Thermowood	22.1	31–33	100.10–100.26
Plywood	22.3	31–33	100.10–100.27
Blockboard	22.3	31–33	100.10–100.23

4.3 Results

4.3.1 Plastics

The time to ignition recorded for the various plastics for the various individual external heat fluxes is shown in Fig. 4.2. As we can see the values for the PLA and ABS-T samples are similar. The difference in the time to ignition between them was 5.5 s for a heat flux of 55 kW m^{-2} and up to 18 s at 25 kW m^{-2}. The time to ignition of PLA was longer than that of ABS. The PETG samples behaved quite differently. While the clear filament required a substantially longer time to ignite than PLA and ABS, PETG with a fire retardant ignited in a shorter time. With decreasing heat flux, the differences in time to ignition between the individual samples increased.

Thermal Thickness

The thermal thickness of plastics, in the context of their autoignition, may be calculated based on the relationship shown in (3.21). Since one of the input values, apart from the external heat flux, is density or volume weight, the thermal penetration depth may be calculated for both materials (Fig. 4.3) and for the samples themselves (Fig. 4.4).

As the indicated correlations clearly demonstrate, the thermal penetration depth for autoignition should in all cases be less than the thickness of samples, which is 8 mm. Since the autoignition time is longer when only an external heat flux is applied in comparison with ignition when a pilot spark is present, the plastic samples may also behave as if they were thermally thick.

It is rather difficult to determine the thermal thickness of plastic samples printed by a 3D printer, in particular due to the use of filling layers that create relatively large empty spaces in the sample. Therefore, the heat transfer on the surface and

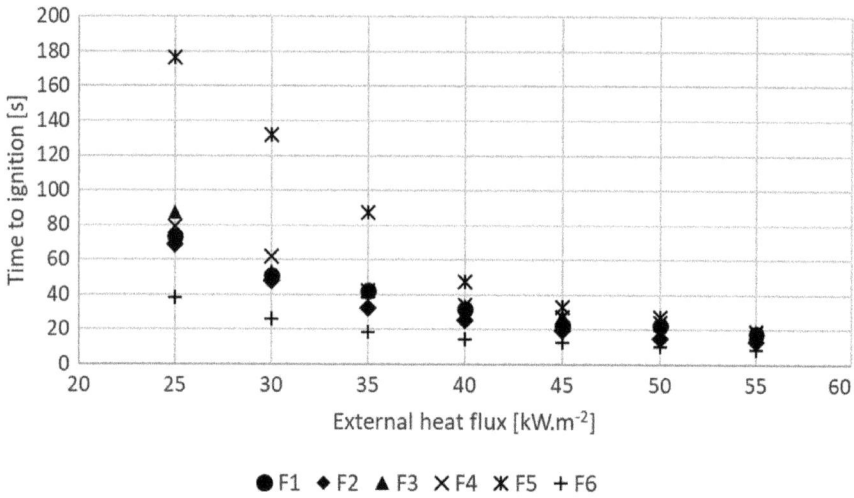

Fig. 4.2 Correlation between time to ignition and external heat flux seen from the measurements of synthetic polymers

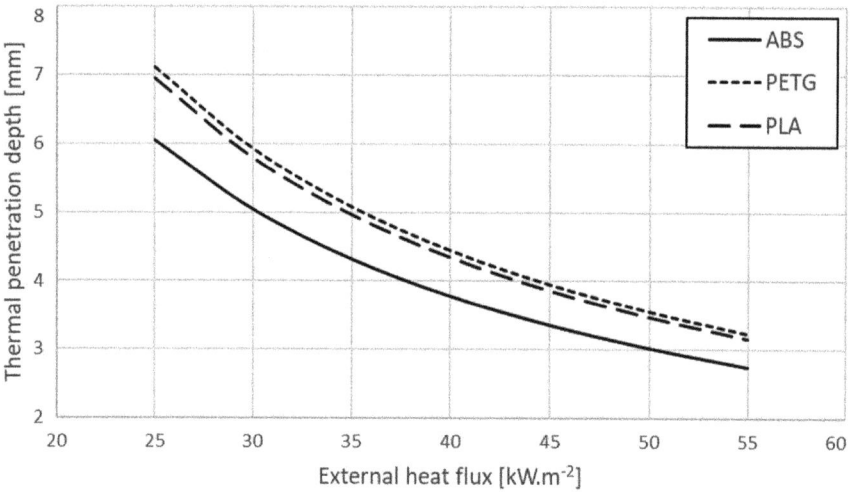

Fig. 4.3 Thermal penetration depth for the autoignition of the tested materials

in the middle of the sample may significantly differ. Temperature measurements were conducted during the application of the external heat flux in order to classify the thermal behaviour of the samples. The heat flux was set to 25 kW m^{-2}. The temperatures of the upper and bottom surface layers of the sample were measured using K-type thermocouples. The results of these measurements are shown in Fig. 4.5.

In the case of ABS-T and PLA, there was a significant difference between the temperature of the upper and bottom surface layers at the time of ignition. However,

4.3 Results

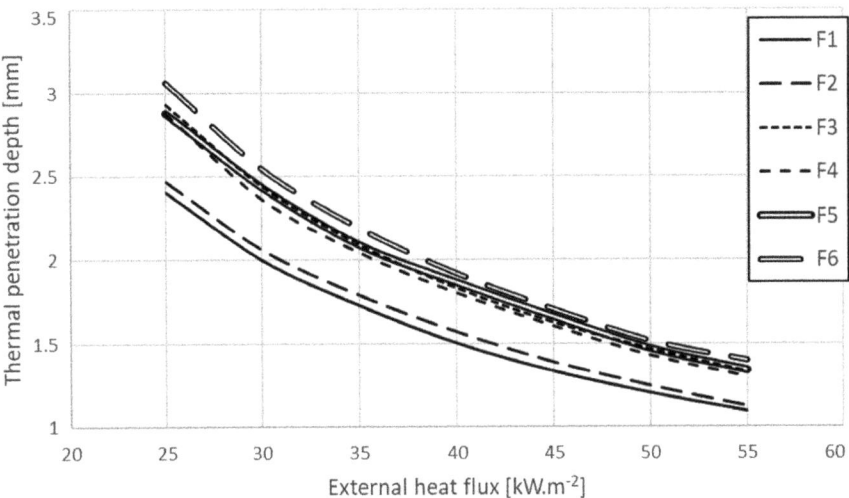

Fig. 4.4 Thermal penetration depth for the autoignition of the tested samples

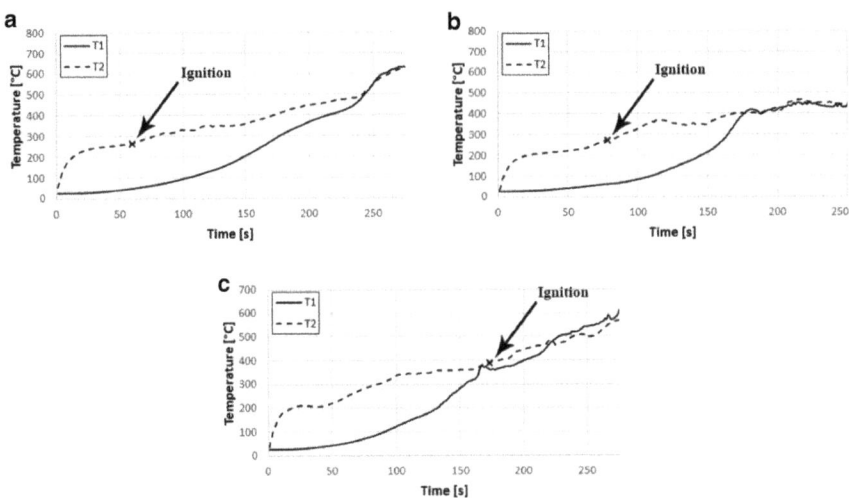

Fig. 4.5 Comparison of the temperatures of the lower (T1) and upper (T2) surfaces of the samples at an external heat flux of 25 kW m^{-2}: **a** ABS-T; **b** PLA; **c** PETG

the temperature difference in the case of PETG is not significant, and ignition takes place approximately 5 s after the upper and lower temperatures are equal. As the external heat flux decreases, the thermal penetration depth increases. Thus, at higher values of external heat flux, these materials appear to be increasingly more thermally thick. Based on the graphs shown in Fig. 4.5, the ABS-T and PLA samples should behave as thermally thick materials. On the other hand, the PETG samples should react to lower external heat fluxes as thermally thin materials and to higher heat fluxes as thermally thick materials. But the 3D print method may make such assumptions somewhat complicated, since the interior of the samples is filled by a structure of reinforcing patterns (in this case a cubed pattern). On the other hand, the surface is created by a compact layer. This leads us to the question whether it is possible to determine the thermal thickness of the whole sample, of its upper layer, or if it is partially determined by the characteristics of the surface and partially by the characteristics of the internal filler.

Critical Heat Flux

Using Table 3.1 from "Methods of Calculation of Ignition Parameters", n-coefficient values were determined, and subsequently, the respective correlations were calculated. These were used for the calculations of critical heat flux. The resultant values along with the n-exponents and the coefficient of determining the trend line are indicated in Tables 4.4, 4.5, and 4.6. Since none of the materials tend to self-ignite under normal circumstances, all negative values of critical heat flux may be excluded. It is equally possible to exclude values higher than 25 kW m^{-2}, since it was proven experimentally that the samples ignited at this level of external heat flux.

Table 4.4 Critical heat flux of ABS-T samples

n	F1		F2	
	R^2	q_{cr}	R^2	q_{cr}
1	0.9752	16.4	0.9840	19.7
1.25	0.9795	11.2	0.9908	15.4
1.5	0.9810	5.9	0.9935	11.1
1.83	0.9814	−1.1	0.9945	5.3
2	0.9813	−4.7	0.9945	2.3
2.8	0.9801	−21.8	0.9931	−12.0
3	0.9797	−26.1	0.9927	−15.5

Table 4.5 Critical heat flux of PLA samples

n	F3		F4	
	R^2	q_{cr}	R^2	q_{cr}
1	0.9907	16.7	0.9827	17.0
1.25	0.9948	11.6	0.9895	11.9
1.5	0.9951	6.4	0.9922	6.7
1.83	0.9936	−0.4	0.9933	−0.1
2	0.9926	−3.9	0.9934	−3.7
2.8	0.9878	−20.6	0.9925	−20.4
3	0.9868	−24.8	0.9921	−24.7

4.3 Results

Table 4.6 Critical heat flux of PETG samples

n	F5		F6	
	R^2	q_{cr}	R^2	q_{cr}
1	0.9434	25.0	0.9932	16.9
1.25	0.9633	22.1	0.9965	11.9
1.5	0.9738	19.1	0.9967	6.8
1.83	0.9810	15.0	0.9953	0.1
2	0.9832	12.9	0.9944	−3.4
2.8	0.9878	2.9	0.9902	−19.9
3	0.9883	0.4	0.9893	−24.0

It is also possible to estimate the approximate values of critical heat flux based on the measured temperature of the upper surface of the sample at the time of ignition of flaming combustion. Based on data reported by Försth and Roos [6], the mean absorptivity of plastics under an external heat flux of 25 kW m^{-2} may equal 0.91. Formula (3.23) should then look as follows:

$$q_{cr} = \frac{T_{ig}^4}{1.602 \times 10^7} \qquad (4.2)$$

Consequently, it is possible to determine an estimated value of the n-coefficient based on the measurement data and calculated critical heat flux. The values thus obtained are listed in Table 4.7.

The appropriate correlations were finally determined as the closest n-values in comparison with the data obtained by calculation based on the measured ignition temperature.

Flux–Time Product

The flux–time product was calculated based on the slope of the trend line of the correlation between the external heat flux reaching the surface of the samples and the inverse value of the time to ignition raised to the power of n. The FTPs of the 3D filaments are summarised in Table 4.8. The value of FTP is always substantially

Table 4.7 Critical heat fluxes determined from the ignition temperatures and corresponding n-coefficients of synthetic polymer samples

Filament	q_{cr} [kW m^{-2}]	n [−]
F1	5.4	1.52
F2	5.8	1.80
F3	5.6	1.54
F4	5.8	1.54
F5	12.1	2.06
F6	4.7	1.60

Table 4.8 Flux–time products for the individual measured filaments

Filament	n	FTP [kW s$^{1/n}$ m^{-2}]	FTP^n [kW^n s m^{-2n}]
F1	1.5	5.997	330.1
F2	1.83	16.251	200.0
F3	1.5	6.430	345.8
F4	1.5	6.464	347.0
F5	2	33.856	184.0
F6	1.5	2.841	200.6

affected by the n-coefficient. Hence, it is better to quote it in the FTP form, which is appropriate for use in calculations of time to ignition.

Thermal Inertia and Thermal Response Parameter

None of the studied 3D polymer samples behaved as thermally thick material. While the measurements taken of the PETG surface temperatures were very similar on the upper and lower surfaces (at an external heat flux of 25 kW m^{-2}), the remaining samples cannot be considered to be thermally thick due to their low n-coefficients. This leads to problems in the calculations of both TRP and the thermal inertia related to the time to ignition of the filaments. They can be determined in part based on the known values of density, thermal conductivity, and thermal capacity of the individual polymers. But even so complications resulting from the printing method still occur.

Sonsalla et al. [7] suggest that the higher density of the filler improves the thermal conductivity. At the same time, the implementation of a thicker layer height marginally increased thermal conductivity. The impact of printing speed on thermal conductivity is negligible.

The values of thermal inertia and the thermal response parameter of individual materials may be determined based on the values reported in literature. However, these values do not relate to the condition at the moment of ignition. Therefore, they should be considered as approximate values. An overview of thermal conductivity, density, and thermal capacities of ABS, PLA, and PETG is given in Table 4.9.

Since the polymer materials used melted before they started to burn, the thermal response parameter as well as thermal inertia may be more accurately determined based on their melting data. Unfortunately, this data is not always readily available. Example values for ABS are given in Table 4.9.

In addition to the data in Table 4.9, the values necessary for the calculation of the thermal response parameter include the difference between the ignition temperature and the ambient temperature. For this purpose, the selected temperatures for the individual materials were obtained using thermocouples (Table 4.10).

Ignition Temperature

The ignition temperature was determined in three ways. The first method involved approximate measurements using a thermocouple. However, this method only provides an approximate value. Polymer samples start to soften and melt when exposed to an external heat flux. The placement of the thermocouple on the sample

4.3 Results

Table 4.9 Thermal conductivity, density, and thermal capacity of selected synthetic polymers

	K [W m^{-1} K^{-1}]	ρ [kg m^{-3}]	c_p [J kg^{-1} K^{-1}]	Source
ABS	0.155[a]	960[a]	2.350[a]	[8]
ABS	0.177	1.050	2.080	[9]
ABS	0.24	1.060–1.080	1.260–1.675	[10]
ABS	0.197	1.039	1.280	[11]
ABS	0.28		1.300	[12]
PMMA	0.193	1.184	1.450	[10]
PLA	0.13	1.240	1.800	[13]
PLA	0.12–0.15	1.210–1.240	1.180–1.210	[14]
PLA	0.1208	1.250	1.624	[15]
PLA	0.081	1.225	1.208.4	[16]
PETG	0.29	1.290	1.200	[17]
PETG	0.2	1.270	–	[18]
PETG	0.21	1.270	1.300 (60 °C) 1.760 (100 °C) 1.880 (150 °C) 2.050 (250 °C)	[19]
PETG	0.225	1.270	1.171	[20]

[a] ABS melt

Table 4.10 Thermal inertia, thermal diffusivity, and the thermal response parameter of selected polymers

Material	I [kW^2 s m^{-4} K^{-2}]	κ [m^2 s^{-1} × 10^7]	TRP [kW s$^{0.5}$ m^{-2}]
ABS (melt)	0.350	0.69	150.3
ABS	0.361	1.39	152.6
PLA	0.210	0.65	117.5
PETG	0.361	1.48	221.7

surface may therefore partially change when attached in a horizontal orientation, which may result in recording a temperature lower than the actual surface temperature.

The second method involved the determination of the temperature using a black body assumption. The calculations were made according to Formula 3.22. In comparison with the actual measured values, such calculated values tend to be higher than the actual temperature as in reality a grey object reflects some of the energy into the surrounding area.

The third calculation method was based on Formula 3.24. Since the samples were measured in a horizontal orientation, the value of h_c was set to 5 W m^{-2} K^{-1}

Table 4.11 Ignition temperatures of samples of synthetic polymers

Filament	T_{ig} [°C]		
	Measured	Equation 3.22 (T_{bb})	Equation 3.24
F1	270	295	260
F2	280	280	245
F3	275	308	274
F4	280	314	280
F5	390	418	389
F6	250	316	283

(Delichatsios [63]). The emissivity and absorptivity were considered constant for all samples. The utilised value was 0.92, which, according to Försth and Roos [6], is the mean value for plastics exposed to an external heat flux of 10 kW m^{-2}. This calculation could potentially provide the most accurate ignition temperature compared to the previous methods, but it is more difficult to use due to its complexity and the amount of data required.

An overview of the ignition temperatures obtained through the three methods outlined above is provided in Table 4.11.

4.3.2 Wood-Based Materials

As was done for plastics, the results for wood-based materials were used to determine the correlation between the time to ignition and an external heat flux (Fig. 4.6). Generally, thermowood and pine had shorter times to ignition, but this is not necessarily a rule.

Thermal Thickness

The homogeneity of wood-based materials is very low in comparison with synthetic polymers. Hence, in order to determine ignition characteristics, it is necessary to conduct a large number of measurements or know the correlation between time to ignition and applied heat flux with sufficient accuracy. As already described in "Methods of Calculation of Ignition Parameters", many authors made experimental measurements on wood or its composites. Their thermal penetration depth was calculated using Formula 3.19 based on the density of the individual samples and the external heat flux applied. The results are given in Table 4.12. As already mentioned above, all of the samples were thermally thick.

The classification of the thermal thickness of wood is even clearer if it is based on the temperatures measured on the top and bottom surfaces as shown in Fig. 4.7. Since wood does not melt, the measurement of its surface temperature is considerably more accurate than in the case of plastics.

4.3 Results

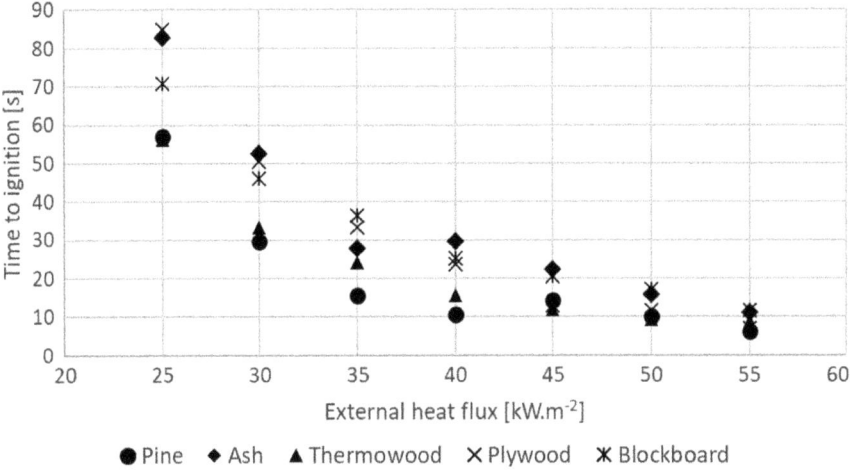

Fig. 4.6 Correlation of time to ignition and external heat flux during measurements of wood-based materials

Table 4.12 Thermal thickness of the analysed wood-based materials

Material	δ [mm]	L_0 [mm]	Thermal thickness
Pine	4.99–11.15	18	Thermally thick
Ash	7.53–16.06	18–20	Thermally thick
Thermowood	4.93–10.71	19	Thermally thick
Plywood	7.33–16.12	18–18.5	Thermally thick
Blockboard	6.18–14.47	17–17.5	Thermally thick

Since all samples may be considered thermally thick, in order to determine their critical heat flux, it would be ideal to use a correlation with an n-coefficient of 2. The second option is to use the calculation from [21], which was also designed for thermally thick materials and is a viable option for wood, but it uses a slightly lower n-coefficient of 1.83.

The correlation between external heat flux and the time to ignition to the power of n was used to determine intercept heat fluxes that were subsequently recalculated using Formula 3.3 to critical heat flux values (Table 4.13).

The appropriate n-value may also be calculated based on the measured data and the critical heat flux corresponding to the ignition temperature measured (Table 4.14).

Flux–Time Product

Just as was done for synthetic polymer filaments, the flux–time product of wood was also calculated based on the slope of the trend line of the corresponding correlation. Since there were two possible n values for wood, the FTP was calculated for both of them. The results obtained are given in Table 4.15.

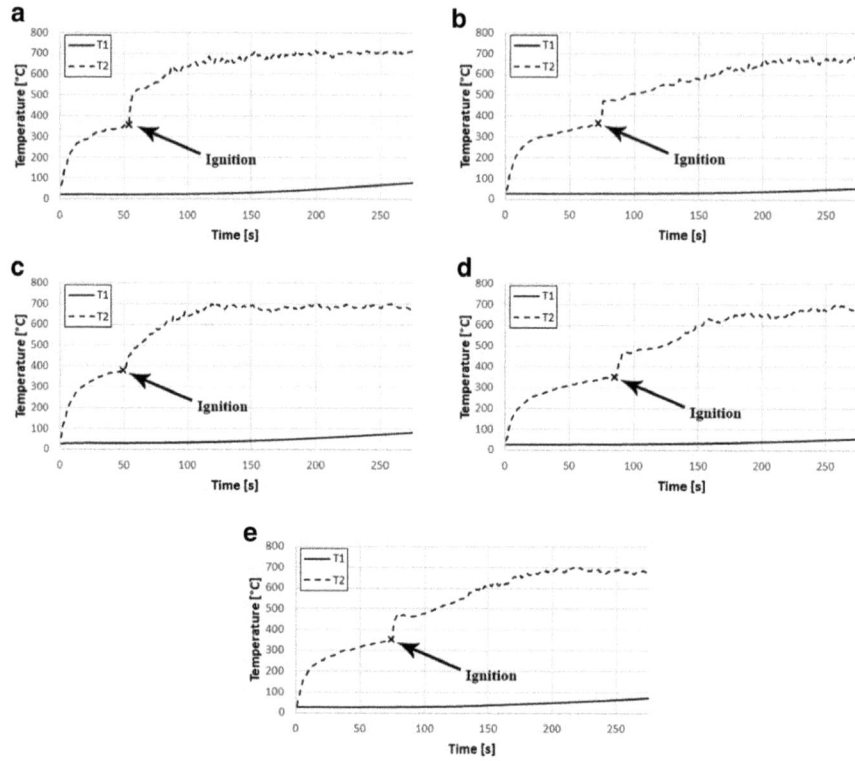

Fig. 4.7 Correlation over time between the temperature of the top and bottom surfaces of the sample at an external heat flux of 25 kW m^{-2}: **a** pine, **b** ash, **c** thermowood, **d** plywood, and **e** blockboard

Table 4.13 Critical heat fluxes and determination coefficients for samples of wood-based materials

Material	$n = 1.83$		$n = 2$	
	R^2 [–]	q_{cr} [kW m^{-2}]	R^2 [–]	q_{cr} [kW m^{-2}]
Pine wood	0.8882	11.32	0.8891	7.75
Ash wood	0.9545	11.88	0.9567	8.29
Thermowood	0.9878	11.27	0.9868	7.77
Plywood	0.9640	21.39	0.9688	18.71
Blockboard	0.9805	8.73	0.9829	4.80

The Thermal Inertia and Thermal Response Parameter

It has already been explained that the thermal inertia and the thermal response parameter may be calculated using the thermal characteristics of the material or the trend line formula for the measured data. The results of the first option are the values given in Table 4.15, and the data from option two is provided in Table 4.16. The thermal

Table 4.14 Critical heat fluxes determined based on ignition temperatures and the corresponding n-coefficients of synthetic polymer samples

Material	q_{cr} [kW m^{-2}]	n [–]
Pine wood	10.83	1.85
Ash wood	11.14	1.86
Thermowood	12.34	1.77
Plywood	10.27	2.56
Blockboard	10.46	1.75

Table 4.15 Flux–time product of the tested wood-based materials

Material	$n = 1.83$		$n = 2$	
	FTP [kW s$^{1/n}$ m^{-2}]	FTP^n [kW^n s m^{-2n}]	FTP [kW s$^{1/n}$ m^{-2}]	FTP^n [kW^n s m^{-2n}]
Pine wood	6.217	118.3	12.996	114.0
Ash wood	12.489	173.2	26.569	163.0
Thermowood	8.875	143.7	18.796	137.1
Plywood	5.969	115.7	12.254	110.7
Blockboard	15.067	191.9	32.652	180.7

conductivity was determined based on the mean density of the individual materials and the measured ignition temperature at an external heat flux of 25 kW m^{-2}. Formula 2.8 was used to calculate the heat capacity at the ignition temperature. Since the samples were dried before the measurements, the impact of humidity does not have to be taken into consideration. For plywood and blockboard, the thermal conductivity was adjusted using a coefficient of 0.86 according to Formula 2.14.

When exposed to an external heat flux, wood degrades while generating a carbonated residue. According to Kashiwagi et al. [22], the absorptivity of surface chars is around 0.95. The convection coefficient at ignition was calculated according to Formula 2.52. Two methods were used to calculate the values in Table 4.17. For $n = 1.83$, thermal inertia was determined based on the correlation between external heat flux and the square of the time to ignition, after which the thermal response parameter

Table 4.16 Thermal properties of the samples of wood-based materials

Material	ρ [kg m^{-3}]	K [W m^{-1} K^{-1}]	c_p [kJ kg^{-1} K^{-1}]	κ [m^2 s^{-1} $\times$ 10^7]	I [kW^2 s m^{-4} K^{-2}]
Pine wood	470	0.2338	2.66	1.87	0.2923
Ash wood	676	0.3214	2.68	1.77	0.5825
Thermowood	432	0.2260	2.76	1.90	0.2692
Plywood	647	0.2602	2.62	1.53	0.4414
Blockboard	565	0.2326	2.63	1.56	0.3462

was calculated based on the measured ignition temperature and the initial ambient temperature. For $n = 2$, the opposite process was used, meaning that the thermal response parameter was calculated based on the above-mentioned correlation and the thermal inertia was determined based on the temperature difference.

The Ignition Temperature

As in the case of synthetic polymers, the ignition temperature was determined through three methods. The first was experimental. The temperature of the surfaces exposed to an external heat source was measured during testing using thermocouples. The charts shown in Fig. 4.7 were compiled using the data obtained, which were subsequently used to determine the temperature at the time of ignition.

The second method was based on the assumption that the surface of the wood carbonises when exposed to a thermal load, which produces properties similar to that of a black body. The temperatures that corresponded to the critical heat flux were calculated using Stefan–Boltzmann law (Formula 3.22).

In the third method, the emissivity of the surface of the wood was taken into consideration. As already mentioned, a value of 0.95 was used in the calculations [22]. As in the calculations for synthetic polymers, the values of h_c were set to 5 W m^{-2} K^{-1} (Delichatsios [63]). A comparison of the ignition temperature for all three methods is given in Table 4.18.

4.4 Discussion

A relatively large number of authors have dealt with the measurement of the ignition parameters of polymer materials. A selection of their data is given in Tables 4.19 and 4.20. It is clear that these authors tend to only mention some of the criteria, and this is mostly due to their use, by the authors, to describe the correlation between the time to ignition of a sample and external heat flux. In such cases, the thermal response parameter or thermal inertia is often determined, since these characteristics describe a very similar property of the samples. Some authors only mention critical heat flux and ignition temperature, which are quite easily applied to fire protection.

The process of ignition is highly dependent not only on the physical and chemical characteristics of the measured material, but also on the test conditions. The characteristics of the ignition source and the orientation of the sample have a substantial impact. Generally, the shortest time to ignition results from the use of a flaming ignitor, a longer time is produced when using an ignition spark and the longest times occur when ignition is triggered without an ignitor (i.e. autoignition). For instance, Shi and Chew [23] used a critical heat flux greater than 25 kW m^{-2} when observing the autoignition of six types of wood, which is considerably higher than is commonly used with pilot ignition.

As was already mentioned in "Methods of Calculation of Ignition Parameters", the critical heat flux for vertically orientated samples is approximately 15% higher than for samples tested in the horizontal plane. In addition, for a horizontal sample, there

4.4 Discussion

Table 4.17 Convection coefficients at ignition and the thermal inertia of the tested wood-based materials

Material	$n = 1.83$	$n = 1.83$	$n = 1.83$	$n = 2$	$n = 2$
	Convection coefficient at ignition [W m^{-2} K^{-1}]	I [kW^2 s m^{-4} K^{-2}]	TRP [kW s$^{0.5}$ m^{-2}]	I [kW^2 s m^{-4} K^{-2}]	TRP [kW s$^{0.5}$ m^{-2}]
Pine wood	24.3	0.157	133.5	0.184	144.6
Ash wood	25.2	0.271	178.0	0.316	192.2
Thermowood	22.5	0.160	143.9	0.190	156.8
Plywood	47.2	0.154	128.5	0.154	128.9
Blockboard	19.1	0.315	185.7	0.393	207.4

Table 4.18 Ignition temperatures of samples of wood-based materials

Material	T_{ig} [°C]				
	Measured	$n = 1.83$		$n = 2$	
		Equation 3.22 (T_{bb})	Equation 3.24	Equation 3.22 (T_{bb})	Equation 3.24
Pine wood	359	396	367	337	305
Ash wood	364	404	376	345	314
Thermowood	382	394	365	336	305
Plywood	350	511	488	484	459
Blockboard	353	354	323	267	233

is a relatively constant temperature distribution across the sample before ignition and thus a shorter time to ignition [24]. Observations of the ignition of vertical samples were also described by Tran and White [25]. At the same time, they used a flaming ignitor. Under these conditions, they determined a critical heat flux for wood from 10.00 to 12.42 kW m^{-2}, an ignition temperature from 298 and 360 °C, and apparent thermal inertia in the range 0.073–0.360 kW^2 s m^{-4} K^{-2}. In this case, it appears that the greater difficulty experienced in the ignition of vertical samples is compensated for by the use of a flaming ignition source.

Olson et al. [26] suggested a further possibility for testing of samples in the horizontal plane with the heater placed under the sample, to allow the simulation of conditions within a space. In this configuration, the influence of buoyancy is reduced because the rise of hot gases is inhibited [27].

The problem when comparing the ignition characteristics of various materials is not only related to the test conditions, but also to their method of calculation. Each of the methods described in the previous text has its specific limitations. Simplifications are often made, which lead to the limits of use of the different methods.

4.4.1 Synthetic Polymers

As previously mentioned above, synthetic polymers are made up of a vast group of macromolecule substances. They include both thermally highly resistant materials as well as materials that can ignite at relatively low temperatures. This leads to a wide range of ignition characteristics. The reported critical heat fluxes range from 4 to 60 kW m^{-2} (Table 4.18). The data in Table 4.18 was used to create a chart of the correlation between critical heat flux and the molar ratio of hydrogen in the corresponding monomer units (Fig. 4.8). It is apparent that as the amount of hydrogen in the molecule increases, the critical heat flux of synthetic polymers decreases.

For all the printable polymers observed, the specified critical heat flux corresponded to lower values when compared to other types of plastics. While the critical heat flux of ABS-T printed samples was significantly lower in comparison with ABS, in the case of PETG, it was very similar to the values obtained for PET. ABS is a

4.4 Discussion

Table 4.19 Ignition characteristics of synthetic polymers as quoted in literature

Material	n [–]	q_{cr} [kW m^{-2}]	I [kW^2 s m^{-4} K^{-2}]	TRP [kW s$^{0.5}$ m^{-2}]	T_{ig} [°C]	Source
ABS		15			394	[31]
PA6		7			432	[31]
PA66		14			456	[31]
PAI		37			526	[31]
PBT		8			382	[31]
PC		17			375–500	[31]
PE		13			380	[31]
PEEK		27			570	[31]
PEI		21			528	[31]
PEN		22			479	[31]
PET		12			407	[31]
PI		33			600	[31]
PMMA		11			280–320	[31]
POM		23			344	[31]
PP		11			330–370	[31]
PPS		37			520–575	[31]
PPSU		41			575	[31]
PS		13			345–370	[31]
CPVC		60			643	[31]
ETFE		16			540	[31]
FEP		40			630	[31]
PTFE		51			630	[31]
PVC		44			395	[31]
PVDF		44			643	[31]
PMMA	2	4	2.12		180	[28]
EPS	2	11.77–12.07				[32]
XPS	2	10.42–10.82				[32]
Polyester cotton (65:45) fabric	1	20				[33]
Acrylic fabric	1	9				[33]

(continued)

Table 4.19 (continued)

Material	n [-]	q_{cr} [kW m^{-2}]	I [kW^2 s m^{-4} K^{-2}]	TRP [kW s$^{0.5}$ m^{-2}]	T_{ig} [°C]	Source
HDPE	2	7.9				[34]
PMMA	2	4.8				Luche [29]
PEI	1	35.8				[35]
PES	1	32.2				[35]
PEEK	1	30.8				[35]
PPO	1	17.9				[35]
PVC	1	14.2				[35]
HIPS	1	13.4				[35]
PBT	1	12.8				[35]
PC	1	12.8				[35]
PMMA	2	5			277–287	[30]
PE	2	15		345	443	[36]
PP + 2.2% inert	2	15		240	443	[36]
PP + 0.2% inert	2	15		237	443	[36]
PP + 20.4% inert	2	15		208	443	[36]
PC + 0.2% inert	2	20		252	497	[36]
PVC + 8% inert	2	10		176	357	[36]
PP + 18.8% inert	2	15		238	443	[36]
PVC + 52.9% inert	2	10		214	374	[36]
PET + 1.% inert	2	10		113	374	[36]
PMMA	2	10		259	378	[36]
EPS	2	15	0.91		376	[37]
PEEK		24	0.702		535	[38]

(continued)

4.4 Discussion

Table 4.19 (continued)

Material	n [–]	q_{cr} [kW m^{-2}]	I [kW^2 s m^{-4} K^{-2}]	TRP [kW s$^{0.5}$ m^{-2}]	T_{ig} [°C]	Source
PA6		20.0	0.798		465	[39]

Table 4.20 Ignition characteristics of wood-based materials as quoted in literature

Material	n [–]	q_{cr} [kW m^{-2}]	I [kW^2 s m^{-4} K^{-2}]	TRP [kW s$^{0.5}$ m^{-2}]	T_{ig} [°C]	Source
Fibre insulation board	1.5	6.25	0.012			[46]
Western red cedar	1.5	14.57	0.045			[46]
American whitewood	1.5	14.57	0.078			[46]
Freijo	1.5	14.99	0.108			[46]
African mahogany	1.5	12.49	0.113			[46]
Oak	1.5	12.49	0.127			[46]
Iroko	1.5	12.49	0.175			[46]
Douglas fir	2	18		182	478	[47]
Scots pine	2	19		164	488	[47]
Southern pine	2	19		201	488	[47]
Shorea	2	16		152	456	[47]
Merbau	2	40		275	643	[47]
Redwood	2	15.5	0.22		375	[48]
Red oak	2	10.8	1.01		304	[48]
Douglas fir	2	16.0	0.25		384	[48]
Maple	2	13.9	0.67		354	[48]
Light cotton fabric	1	17				[33]
Heavy cotton fabric	1	10				[33]
Heavy silk fabric	1	12				[33]
Wool fabric	1	11				[33]
Australian radiata pine	2	18		201	478	[49]
Laminated bamboo	2	7–8		269–376	320–340	[50]
Fir	2	11.6	0.25	136	302.2	[51]
Pine needles	2	12	0.15	135	366	[52]
Soybean straw	1	25.31				[53]
Peanut straw	1	29.14				[53]

(continued)

Table 4.20 (continued)

Material	n [–]	q_{cr} [kW m^{-2}]	I [kW^2 s m^{-4} K^{-2}]	TRP [kW s$^{0.5}$ m^{-2}]	T_{ig} [°C]	Source
Rape straw	1	18.27				[53]
Plywood	2	13.4–16.5	0.139–0.326	225.6–230.2		[54]
Nordic spruce	2	19	0.14	291	488	[55]
Leadwood	2	15.0	11.5	376.2	149	[45]
Mopani	2	14.4	10.6	161.2	77	[45]
Tamboti	2	5.9	5.8	352.7	187	[45]
Stinkwood	2	9.2	5.9	173.6	102	[45]
Real Yellowwood	2	1.3	2.5	232.2	187	[45]
Western cedar	1.83	13.3	0.087		354	[56]
Redwood	1.83	14.0	0.141		364	[56]
Radiata pine	1.83	12.9	0.156		349	[56]
Douglas fir	1.83	13.0	0.158		350	[56]
Victorian ash	1.83	10.4	0.260		311	[56]
Blackbutt	1.83	9.7	0.393		300	[56]
Spruce	1.83	14.1	0.181		352	[57]
Poplar	1.83	14.5	0.101		356	[57]
Oak	1.83	10.6	0.447		301	[57]
Beech	1.83	7.5	0.783		246	[57]
Softwood	2.2	10.0				[58]
Chipboard	1.7	9.0				[58]
Plywood	1.5	8.5				[58]

copolymer of several monomers. Its properties may differ depending on the representation of the individual monomers in the mix. Moreover, the filaments were made of ABS-T, which means that the initial material was enriched with a small amount of PMMA. For PMMA, some authors suggest critical heat flux values ranging from 4 to 5 kW m^{-2} [28–30], while others suggest 10 kW m^{-2} and more (Tewarson et al. [64]) [31]. PLA, which is a fundamental printing polymer, is highly sensitive to the increases in temperature. Thus, the lower value of critical heat flux is not surprising.

Other ignition characteristics are less often described in the literature in comparison with critical heat flux. The thermal inertia of polymers is quoted in the range of values from 0.702 to 2.12 (Table 4.19); however, considering the small number of references, this cannot be applied as a general rule. The same is true for the thermal response parameter. Tewarson et al. [64] suggest that it is in the range of 113–345 kW s$^{0.5}$ m^{-2}.

Ignition temperature is more often stated by authors. For the various polymers shown, it may differ considerably, with values ranging from 180 to 643 °C (Table 4.19).

4.4 Discussion

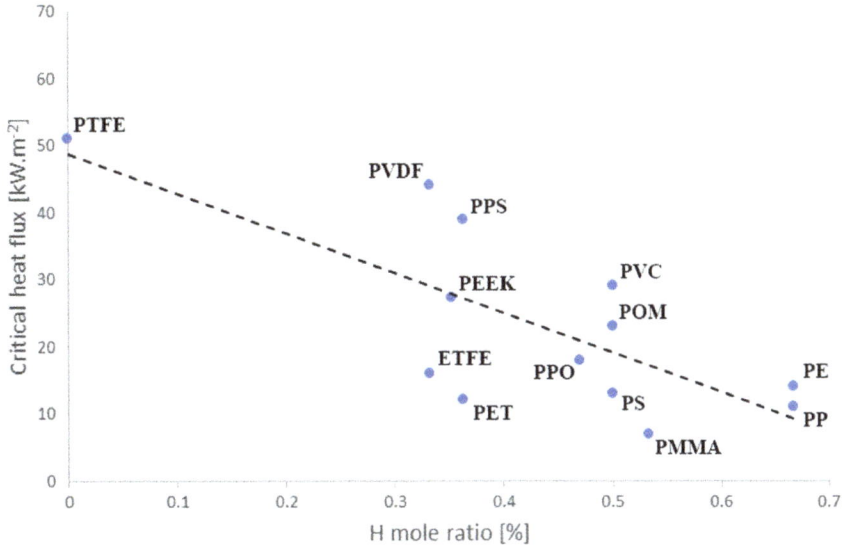

Fig. 4.8 Correlation between critical heat flux and the amount of hydrogen in the molecule of synthetic polymers

Synthetic polymers also tend to be mixed with various additives. These additives, in addition to the effect they have on other characteristics, also influence the process of ignition. Nelson et al. [40] suggest that based on their measurements, inert additives increase the critical heat flux and the time to ignition, however if the sum of the thermal capacity and density of the additive is lower than that of the polymer, the time to ignition may even decrease. The presence of dispersed nanoclays in polymers may have a strong impact on the ignition processes of polymers, both in terms of a reduction in the time to ignition and the thickness of the material that contributes to fuel production at ignition [41].

Kashiwagi and Cleary [42] observed the effect of mounting on the flammability properties of intumescent polymers. For PEI, they measured a critical heat flux of 28–30 kW m^{-2}, ignition temperature from 507 to 524 °C, and thermal inertia from 2.45 to 3.36 kW^2 s m^{-4} K^{-2}. For PC, they measured a critical heat flux of 22–23 kW m^{-2}, ignition temperature from 455 to 464 °C, and thermal inertia from 1.75 to 1.76 kW^2 s m^{-4} K^{-2}. However, it must be mentioned that they used a horizontally oriented lateral flame spread apparatus for the tests with an air acetylene pilot flame as the ignition source.

The temperature of an inert surface rises with increasing external heat flux. In the case of flammable materials, it may be decreased by pyrolysis, degradation, or vaporisation. The pyrolysis temperature of non-charring materials equals the surface temperature [43].

4.4.2 Natural Polymer Materials

Natural polymers include a relatively wide range of materials. Considering that these polymers often make up parts of the bodies of plants and animals, they tend to occur as mixtures in which each macromolecule has a specific function. Their exact composition depends on the conditions under which they are created. For example, the chemical composition of wood differs depending on the part of the tree, the type of tree, its geographic location, the climate, and the properties of the soil in which it grew [44].

Given the values suggested by various authors (Table 4.20), it may be stated that the critical heat flux of natural polymers is in the range of 1.3–40 kW m^{-2}, with the average around 14 kW m^{-2}. The values obtained for wood and wood composites, for the purposes of this monograph, are within this range, slightly below the average.

The range of thermal inertias of natural polymer materials is, from the results of various authors, 0.012–11.5 kW^2 s m^{-4} K^{-2}. However, it should be noted that values above 1.01 kW^2 s m^{-4} K^{-2} have only been reported by Maake et al. [45], who studied South African hardwoods. Wood species that grow in mild climates typically have values around 0.250 kW^2 s m^{-4} K^{-2}, which corresponds to the results determined from the measured data.

When authors used the thermal response parameter for describing ignition, its values for natural polymer materials ranged from 135 to 376.2 kW s$^{0.5}$ m^{-2}. Considering the remaining observed ignition parameters, this is a relatively small range. While for $n = 1.83$, the TRP values are low, and plywood and pine are almost outside the range, with $n = 2$, the TRP values are higher and in each case in the range reported in literature.

As Table 4.20 shows, the ignition temperature of natural polymers ranges from 77 to 643 °C. Although this range is rather broad, only a few authors report values below 200 °C and above 500 °C. The most common temperature range for wood is 300 °C and 400 °C, which is in line with the data in Table 4.20.

At low heat fluxes, the change of surface temperature over time reaches a plateau prior to ignition. The oxidation of the surface layer of char may also have a significant impact on the ignition process (Li and Drysdale [65]).

As a consequence of being a perfectly heterogeneous material, wood has different characteristics in different orientations. Most authors have determined the ignition characteristics along the length of the sample; however, there have been several studies that tested across the sample. Spearpoint and Quintiere [48] suggest that the factors that affect the ignition of wood may also include the species, grain orientation, moisture content, exposure conditions, and inherent variability. Their results show that if the sample under test is oriented along the grain, the critical heat flux and ignition temperature are higher, and on the other hand, the apparent thermal inertia and thermal conductivity are lower. Similar results were reported by Xu et al. [50] for bamboo. Both critical heat fluxes and thermal response parameters and ignition temperatures are higher when the sample is oriented along the grain.

In the case of ignition without an ignitor, we speak of autoignition. Shi and Chew [23] state that in this case, the ignition temperature is about 264–558 °C. Ji et al. [59] suggest surface temperatures that range from 412 to 550 °C for various wood types at the time of ignition, however, no ignitor was used during the measurements and the heat flux applied to the sample increased with time.

The moisture content of the tested samples may also have an impact on the ignition characteristics of natural polymers. Atreya and Abu-Zaid [66] described that with pilot ignition the ignition time increases as the moisture content increases, as does the surface temperature at ignition. But the minimum heat flux required for ignition is approximately the same regardless of moisture content. According to Shi and Chew [60], a higher moisture content results in an increase in ignition temperature with pilot ignition, but no obvious trend in ignition temperature was observed for autoignition when the moisture content was increased from 0 to 11%.

When an external heat flux is applied to natural polymers, not only flaming (homogeneous) combustion, but also heterogeneous combustion may occur. Gratkowski et al. [61] observed this phenomenon on samples made of plywood. Through experiments, they discovered a minimum required heat flux for smouldering ignition of 7.5 kW m^{-2}.

Altitude may also have a significant impact on the ignition of samples. According to Yafei et al. [62], the atmospheric pressure has a major impact on critical heat flux when the remaining conditions are unchanged. Therefore for the ignition characteristics, in addition to other test conditions, it is equally vital to record the atmospheric pressure.

References

1. Gorokhov G, Katsemba M, Liubimau A, Lobko A, Melnikau A (2018) Specifics of 3D-printed electronics. In: International conference on engineering of scintillation materials and radiation technologies. Springer, Cham, pp 315–326
2. Time B (1998) Hygroscopic moisture transport in wood. Norwegian University of Science and Technology, Trondheim
3. Plötze M, Niemz P (2011) Porosity and pore size distribution of different wood types as determined by mercury intrusion porosimetry. Eur J Wood Wood Prod 69(4):649–657
4. Ling Z, Ji Z, Ding D, Cao J, Xu F (2016) Microstructural and topochemical characterization of thermally modified poplar (Populus cathayana) cell wall. BioResources 11(1):786–799
5. Wu Y, Wu X, Yang F, Zhang H, Feng X, Zhang J (2020) Effect of thermal modification on the nano-mechanical properties of the wood cell wall and waterborne polyacrylic coating. Forests 11(12):1247
6. Försth M, Roos A (2011) Absorptivity and its dependence on heat source temperature and degree of thermal breakdown. Fire Mater 35(5):285–301
7. Sonsalla T, Moore AL, Meng WJ, Radadia AD, Weiss L (2018) 3-D printer settings effects on the thermal conductivity of acrylonitrile butadiene styrene (ABS). Polym Testing 70:389–395
8. PM plasty mladeč (2019) ABS transparent. Material data sheet for ABS transparent
9. Zhou Y, Nyberg T, Xiong G, Liu D (2016) Temperature analysis in the fused deposition modeling process. In: 2016 3rd international conference on information science and control engineering (ICISCE). IEEE, pp 678–682

10. Trhlíková L, Zmeskal O, Psencik P, Florian P (2016) Study of the thermal properties of filaments for 3D printing. In: AIP conference proceedings, vol 1752, no 1. AIP Publishing LLC, p 040027
11. Shemelya C, De La Rosa A, Torrado AR, Yu K, Domanowski J, Bonacuse PJ et al (2017) Anisotropy of thermal conductivity in 3D printed polymer matrix composites for space based cube satellites. Addit Manuf 16:186–196
12. Ghavidel AK, Zadshakoyan M (2018) Comprehensive study of laser cutting effects on the properties of acrylonitrile butadiene styrene. Int J Adv Manuf Technol 97(9):3637–3653
13. Sd3D (b), PLA (Polylactic acid). Technical data sheet
14. Ashby MF, Johnson K (2013) Materials and design: the art and science of material selection in product design. Butterworth-Heinemann
15. Auras R (2002) Poly (lactic acid). In: Encyclopedia of polymer science and technology
16. Barkhad MS, Abu-Jdayil B, Iqbal MZ, Mourad AHI (2020) Thermal insulation using biodegradable poly (lactic acid)/date pit composites. Constr Build Mater 261:120533
17. Sd3D (a), PETG (polyethylene terephthalate copolymer). Technical data sheet
18. The plastic people (2013) Technical data sheet PETG
19. rigid.ink (2016) PETG data sheet
20. Eagle (2017) VIVAK-PETG, 2 p
21. Janssens M (1991) Fundamental thermophysical characteristics of wood and their role in enclosure fire growth. Doctoral dissertation, Ghent University
22. Kashiwagi T, Ohlemiller TJ, Werner K (1987) Effects of external radiant flux and ambient oxygen concentration on nonflaming gasification rates and evolved products of white pine. Combust Flame 69(3):331–345
23. Shi L, Chew MYL (2013) Experimental study of woods under external heat flux by autoignition. J Therm Anal Calorim 111(2):1399–1407
24. Tsai KC (2009) Orientation effect on cone calorimeter test results to assess fire hazard of materials. J Hazard Mater 172(2–3):763–772
25. Tran HC, White RH (1992) Burning rate of solid wood measured in a heat release rate calorimeter. Fire Mater 16(4):197–206
26. Olson SL, Beeson HD, Haas JP, Baas JS (2005) An earth-based equivalent low stretch apparatus for material flammability assessment in microgravity and extraterrestrial environments. Proc Combust Inst 30(2):2335–2343
27. Olson S (2011) Convective heat transfer scaling of ignition delay and burning rate with heat flux and stretch rate in the equivalent low stretch apparatus. In: 10th international solid-state sensors, actuators and microsystems conference (No. E-18427)
28. Rhodes BT, Quintiere JG (1996) Burning rate and flame heat flux for PMMA in a cone calorimeter. Fire Saf J 26(3):221–240
29. Luche J, Rogaume T, Richard F, Guillaume E (2011) Characterization of thermal properties and analysis of combustion behavior of PMMA in a cone calorimeter. Fire Saf J 46(7):451–461
30. Tsai TH, Li MJ, Shih IY, Jih R, Wong SC (2001) Experimental and numerical study of autoignition and pilot ignition of PMMA plates in a cone calorimeter. Combust Flame 124(3):466–480
31. Lyon RE, Quintiere JG (2007) Criteria for piloted ignition of combustible solids. Combust Flame 151(4):551–559
32. An W, Jiang L, Sun J, Liew KM (2015) Correlation analysis of sample thickness, heat flux, and cone calorimetry test data of polystyrene foam. J Therm Anal Calorim 119(1):229–238
33. Nazare S, Kandola B, Horrocks AR (2002) Use of cone calorimetry to quantify the burning hazard of apparel fabrics. Fire Mater 26(4–5):191–199
34. Luche J, Mathis E, Rogaume T, Richard F, Guillaume E (2012) High-density polyethylene thermal degradation and gaseous compound evolution in a cone calorimeter. Fire Saf J 54:24–35
35. Goff LJ (1993) Investigation of polymeric materials using the cone calorimeter. Polym Eng Sci 33(8):497–500
36. Tewarson A, Abu-Isa IA, Cummings DR, LaDue DE (1999) Characterization of the ignition behavior of polymers commonly used in the automotive industry. In: Fire safety science, proceedings of the sixth international symposium, pp 991–1002

References

37. Cleary TG (1992) Flammability characterization with the lift apparatus and the cone calorimeter. Fire Retardant Chemicals Association. Technical and marketing issues impacting the fire safety of building and construction and home furnishings applications. Technomic Publishing Co., Lancaster, PA, pp 99–115
38. Patel P, Hull TR, Lyon RE, Stoliarov SI, Walters RN, Crowley S, Safronava N (2011) Investigation of the thermal decomposition and flammability of PEEK and its carbon and glass-fibre composites. Polym Degrad Stab 96(1):12–22
39. Carvel R, Steinhaus T, Rein G, Torero JL (2011) Determination of the flammability properties of polymeric materials: a novel method. Polym Degrad Stab 96(3):314–319
40. Nelson MI, Brindley J, McIntosh AC (1996) Ignition of thermally thin thermoplastics—the effectiveness of inert additives in reducing flammability. Polym Degrad Stab 54(2–3):255–266
41. Fina A, Camino G (2011) Ignition mechanisms in polymers and polymer nanocomposites. Polym Adv Technol 22(7):1147–1155
42. Kashiwagi T, Cleary TG (1993) Effects of sample mounting on flammability properties of intumescent polymers. Fire Saf J 20(3):203–225
43. Schartel B, Hull TR (2007) Development of fire-retarded materials—interpretation of cone calorimeter data. Fire Mater Int J 31(5):327–354
44. Pettersen RC (1984) The chemical composition of wood. In: The chemistry of solid wood, vol 207, pp 57–126
45. Maake T, Asante J, Mwakikunga B (2020) Fire performance properties of commonly used South African hardwood. J Fire Sci 38(5):415–432
46. Lawson DI, Simms UD (1952) The ignition of wood by radiation. Br J Appl Phys 3(9):288
47. Xu Q, Chen L, Harries KA, Zhang F, Liu Q, Feng J (2015) Combustion and charring properties of five common constructional wood species from cone calorimeter tests. Constr Build Mater 96:416–427
48. Spearpoint MJ, Quintiere JG (2001) Predicting the piloted ignition of wood in the cone calorimeter using an integral model—effect of species, grain orientation and heat flux. Fire Saf J 36(4):391–415
49. Delichatsios M, Paroz B, Bhargava A (2003) Flammability properties for charring materials. Fire Saf J 38(3):219–228
50. Xu Q, Chen L, Harries KA, Li X (2017) Combustion performance of engineered bamboo from cone calorimeter tests. Eur J Wood Wood Prod 75(2):161–173
51. Batiot B, Luche J, Rogaume T (2014) Thermal and chemical analysis of flammability and combustibility of fir wood in cone calorimeter coupled to FTIR apparatus. Fire Mater 38(3):418–431
52. Fateh T, Richard F, Batiot B, Rogaume T, Luche J, Zaida J (2016) Characterization of the burning behavior and gaseous emissions of pine needles in a cone calorimeter–FTIR apparatus. Fire Saf J 82:91–100
53. Xie T, Wei R, Wang Z, Wang J (2020) Comparative analysis of thermal oxidative decomposition and fire characteristics for different straw powders via thermogravimetry and cone calorimetry. Process Saf Environ Prot 134:121–130
54. Fateh T, Rogaume T, Richard F, Luche J, Rousseaux P (2010) Characterization of the thermal degradation of two treated plywoods in a cone calorimeter. In: Proceedings of the sixth international seminar on fire and explosion hazards (FEH6), University of Leeds, UK, pp 11–16
55. Hagen M, Hereid J, Delichatsios MA, Zhang J, Bakirtzis D (2009) Flammability assessment of fire-retarded Nordic Spruce wood using thermogravimetric analyses and cone calorimetry. Fire Saf J 44(8):1053–1066
56. Janssens MARC (1991) A thermal model for piloted ignition of wood including variable thermophysical properties. Fire Saf Sci 3:167–176
57. Grexa O, Horvathova E, Osvald A (1997) Cone calorimeter studies of wood species. In: Proceedings of the Korea Institute of Fire Science and Engineering conference. Korean Institute of Fire Science and Engineering, pp 77–84

58. Shields TJ, Silcock GW, Murray JJ (1994) Evaluating ignition data using the flux time product. Fire Mater 18(4):243–254
59. Ji J, Cheng Y, Yang L, Guo Z, Fan W (2006) An integral model for wood auto-ignition under variable heat flux. J Fire Sci 24(5):413–425
60. Shi L, Chew MYL (2012) Influence of moisture on autoignition of woods in cone calorimeter. J Fire Sci 30(2):158–169
61. Gratkowski MT, Dembsey NA, Beyler CL (2006) Radiant smoldering ignition of plywood. Fire Saf J 41(6):427–443
62. Yafei W, Lizhong Y, Xiaodong Z, Jiakun D, Yupeng Z, Zhihua D (2010) Experiment study of the altitude effects on spontaneous ignition characteristics of wood. Fuel 89(5):1029–1034
63. Delichatsios MA (2000) Ignition times for thermally thick and intermediate conditions in flat and cylindrical geometries. Fire Saf Sci 6:233–244
64. Tewarson A, Abu-isa IA, Cummings DR, Ladue DE (2000) Characterization of the ignition behaviour of polymers commonly used in the automotive industry. Fire Saf Sci 6:991–1002
65. Yudong L, Drysdale D (1992) Measurement of the ignition temperature of wood. Fire Saf Sci 1:25–30
66. Atreya A, Abu-Zaid M (1991) Effect of environmental variables on piloted ignition. Fire Saf Sci 3:177–186

Conclusion

The term "polymers" includes both natural and synthetic materials that are based on repeating monomer units. They can vary quite substantially in their composition. Natural polymers tend to have more complex structures owing to their specific function in living organisms. On the other hand, synthetic polymers are simpler and offer more options for modification through the use of various additives.

The thermal degradation of polymers occurs in two fundamental ways. The first is softening, followed by melting and the cleavage of bonds in the macromolecules. In this type of degradation, a relatively large amount of monomer units are released. The second is degradation through the cleavage of functional groups, resulting in the production of a carbon residue. In this case, cross-linking tends to occur, increasing the carbon content. The main products of polymer combustion are carbon dioxide, carbon monoxide, and water.

When radiant heat reaches the surface of a flammable material, some of the thermal energy is reflected, some is conducted into deeper layers, and some heats the surface layer. It is the process of heating the surface layer that has a significant impact on the ignition of flaming combustion. Time to ignition is therefore a function of the applied heat flux. Many authors have discussed the form of this function, and the result of their studies suggests that it is highly dependent upon other factors such as the presence of an ignition source, the type of material, or its thermal thickness.

The lowest heat flux that is able to ignite flaming combustion is referred to as the critical heat flux. It corresponds to the ignition of flaming combustion after it has been applied for a long time to the surface of the flammable material, with an infinitely long time of application (but not necessarily). As such, the critical heat flux partially characterises the ignition of a material. In addition, ignition characteristics include the thermal response parameter, ignition temperature, thermal inertia, and others.

The ignition characteristics were determined for various polymer materials. Of the synthetic polymers, filaments used in 3D printing were selected. These were used to print 100 mm × 100 mm × 8 mm samples through fused deposition modelling.

Two colour variants of the most commonly used PLA were prepared, along with two ABS-T and PETG with and without a fire retardant. Samples of natural polymers included two types of solid wood (pine as a coniferous wood species and ash as a deciduous species), one type of thermally modified wood (thermopine), and two types of wood composite materials (plywood and blockboard). The samples had dimensions of 100 mm × 100 mm with a thickness of 17–19 mm.

A cone calorimeter was used as the measuring apparatus, with the times to ignition, in all cases, measured at external heat fluxes ranging from 25 to 55 kW m^{-2} with an increment of 5 kW m^{-2}. The ignition of flaming combustion took place in the presence of a spark ignitor. The temperature of the upper and the bottom sides of the samples was also measured at the lowest external heat flux.

Based on the results obtained, we may state that:

- The measured synthetic polymers behaved as thermally medium materials, while the wood-based materials had a sufficient thickness to act as thermally thick;
- The critical heat flux of synthetic polymers ranged from 4.7 to 12.1 kW m^{-2}, all the values for PLA and ABS-T were practically identical (5.4–5.8 kW m^{-2}). For wood-based materials, it is practically impossible to determine an appropriate value of the n-coefficient. As a consequence, the critical heat flux of individual materials can be assumed to be between 4.8 and 21.5 kW m^{-2}. The values for solid wood were practically identical, and the critical heat flux of plywood was significantly higher;
- The flux–time product may be highly variable and greatly depend on the n-coefficient. The square of the FTP for filaments was 184.0–347.0. For wood, the calculated FTP^n was from 110.7 to 191.9;
- The thermal response parameters of synthetic polymers ranged from 117.5 to 221.7 kW s$^{0.5}$ m^{-2}. Similar values were also found for wood-based materials, from 128.5 to 204.4 kW s$^{0.5}$ m^{-2};
- Ignition of the tested synthetic polymers occurred at temperatures from 250 to 400 °C, and it is possible to make an approximate estimate based on Stefan–Boltzmann law for black bodies. On the other hand, the measured ignition temperatures of wood were found to be in a relatively small range, from 350 to 382 °C. Values outside the specified temperature range were also calculated, and, especially in the case of plywood, the calculated and measured values significantly differed.

The results obtained have not only provided useful data for the assessment of polymers from the perspective of fire protection, but have also opened other questions in this area which need further research. Considering the possible variability of 3D printing, it is necessary to focus, for example, on the impact of the thickness of the surface layer or the printing method on the initiation of combustion of the resulting prints. For wood-based materials, research into wood composites that consist of multiple layers of wood glued together appears to be important.

GPSR Compliance
The European Union's (EU) General Product Safety Regulation (GPSR) is a set of rules that requires consumer products to be safe and our obligations to ensure this.

If you have any concerns about our products, you can contact us on

ProductSafety@springernature.com

In case Publisher is established outside the EU, the EU authorized representative is:

Springer Nature Customer Service Center GmbH
Europaplatz 3
69115 Heidelberg, Germany

www.ingramcontent.com/pod-product-compliance
Ingram Content Group UK Ltd.
Pitfield, Milton Keynes, MK11 3LW, UK
UKHW021250180426
11946UKWH00003B/56